Direct Steerable Pipe Thrusting

Other Titles of Interest

Pipeline Design for Installation by Horizontal Directional Drilling, Second Edition, edited by Eric R. Skonberg and Tennyson M. Muindi (ASCE/Pipelines Division 2014). MOP 108 addresses the design of major pipeline or duct segments to be installed by horizontal directional drilling (HDD) (ISBN 978-0-7844-1350-0).

Pipe Ramming, Second Edition, by the Pipe Ramming Task Force of the Trenchless Installation of Pipelines (TIP) Committee; edited by Glenn M. Boyce (ASCE/UESI 2020). MOP 115 presents the latest and best practices for the design and construction of road and railroad crossings using pipe ramming technologies (ISBN 978-0-7844-1560-3).

Pilot Tube and Other Guided Boring Methods, by the Task Committee on Pilot Tube and Other Guided Boring Methods (ASCE/UESI 2017). MOP 133 provides a detailed description of the pilot tube and guided boring methods with chapters on project planning, site and geotechnical assessment, shaft design, pipe characteristics and design, contract documents, and construction aspects (ISBN 978-0-7844-1474-3).

Standard Design and Construction Guidelines for Microtunneling, Standard ASCE/CI 36-15. This standard covers the planning, design, materials, and construction for this trenchless method of installing pipelines (ISBN 978-0-7844-1363-0).

Horizontal Auger Boring Projects, by the Horizontal Auger Boring Task Force (ASCE/UESI 2017). MOP 106 covers the horizontal auger boring method, providing the instructions for a safe, productive, and efficient installation of pipelines for road crossings (ISBN 978-0-7844-1458-3).

ASCE Manuals and Reports on Engineering Practice No. 155

Direct Steerable Pipe Thrusting

Sponsored by the
Task Committee on Direct Steerable Pipe Thrusting of the
Committee on Trenchless Installation of Pipelines of the
Utility Engineering & Surveying Institute of the
American Society of Civil Engineers

Published by the American Society of Civil Engineers

Library of Congress Cataloging-in-Publication Data

Names: Robison, Jon, author.
Title: Direct steerable pipe thrusting / Jon Robison, Jeff Scholl, Andrew Sparks, Joachim Engelhardt, Andrew Finney, Norm Joyal, Chris Lamont.
Description: Reston, Virginia : American Society of Civil Engineers, [2023] | Series: ASCE manuals and reports on engineering practice ; no. 155 | Includes bibliographical references and index. | Summary: "MOP 155 details how engineers and construction professionals can utilize the Direct Steerable Pipe Thrusting (DSPT) method to design and install pipelines"-- Provided by publisher.
Identifiers: LCCN 2022059192 | ISBN 9780784416105 (print) | ISBN 9780784484630 (pdf) | ISBN 9780784484647 (epub)
Subjects: LCSH: Underground pipelines--Design and construction. | Pipe-laying machinery. | Trenchless construction. | Boring. | Directional drilling.
Classification: LCC TA660.P55 R63 2023 | DDC 621.8/672--dc23/eng/20230109
LC record available at https://lccn.loc.gov/2022059192

Published by American Society of Civil Engineers
1801 Alexander Bell Drive
Reston, Virginia 20191-4382
www.asce.org/bookstore | ascelibrary.org

***Errata:* Errata, if any, can be found at https://doi.org/10.1061/9780784416105.**

ISBN 978-0-7844-1610-5 (print)
ISBN 978-0-7844-8463-0 (PDF)
ISBN 978-0-7844-8464-7 (ePub)

Manufactured in the United States of America.
28 27 26 25 24 23 1 2 3 4 5

Cover photograph courtesy of Michels Construction.

MANUALS AND REPORTS ON ENGINEERING PRACTICE

(As developed by the ASCE Technical Procedures Committee, July 1930, and revised March 1935, February 1962, and April 1982)

A manual or report in this series consists of an orderly presentation of facts on a particular subject, supplemented by an analysis of limitations and applications of these facts. It contains information useful to the average engineer in his or her everyday work, rather than findings that may be useful only occasionally or rarely. It is not in any sense a "standard," however, nor is it so elementary or so conclusive as to provide a "rule of thumb" for nonengineers.

Furthermore, material in this series, in distinction from a paper (which expresses only one person's observations or opinions), is the work of a committee or group selected to assemble and express information on a specific topic. As often as practicable, the committee is under the direction of one or more of the Technical Divisions and Councils, and the product evolved has been subjected to review by the Executive Committee of the Division or Council. As a step in the process of this review, proposed manuscripts are often brought before the members of the Technical Divisions and Councils for comment, which may serve as the basis for improvement. When published, each manual shows the names of the committees by which it was compiled and indicates clearly the several processes through which it has passed in review, so that its merit may be definitely understood.

In February 1962 (and revised in April 1982), the Board of Direction voted to establish a series titled, "Manuals and Reports on Engineering Practice" to include the manuals published and authorized to date, future Manuals of Professional Practice, and Reports on Engineering Practice. All such manual or report material of the Society would have been refereed in a manner approved by the Board Committee on Publications and would be bound, with applicable discussion, in books similar to past manuals. Numbering would be consecutive and would be a continuation of present manual numbers. In some cases of joint committee reports, bypassing of journal publications may be authorized.

A list of available Manuals of Practice can be found at https://ascelibrary.org/page/books/s-mop.

CONTENTS

PREFACE

This manual of practice was prepared by the Task Committee on Direct Steerable Pipe Thrusting (DSPT) of the ASCE Committee on Trenchless Installation of Pipelines (TIPS), as part of the Utility Engineering & Surveying Institute (UESI). The manual describes the current Direct Steerable Pipe Thrusting method used by engineers and construction professionals in designing and installing pipelines using the DSPT method.

This manual of practice has been created by a group of engineers, contractors, equipment manufacturers, pipeline owners, and academics fully knowledgeable of the method and its use. This manual considers many of the advances that have occurred since the initial development of the technology. However, the task committee acknowledges that DSPT continues to evolve and contractor and equipment capabilities continue to develop.

Sections of this manual of practice have been written assuming that the reader may be unfamiliar with DSPT methodology. No document can encompass all issues on a particular DSPT project. Improvements in best practices and technology continue to advance so that future use of this manual on any project must consider not only the specific characteristics of the particular project but also further improvements in best practices and technology.

The stakeholders of a pipeline trenchless project are encouraged to consider all trenchless methods before concluding that the DSPT method is the most suitable construction method available. Manuals and reports on engineering practice have been written by ASCE and are available for different construction methods. If the engineer responsible for the pipeline project does not have a strong background in trenchless design, an engineering firm that specializes in trenchless designs should be consulted.

ACKNOWLEDGMENTS

The American Society of Civil Engineers (ASCE) and the Utility Engineering & Surveying Institute (UESI) acknowledge the work of the Task Committee on Direct Steerable Pipe Thrusting. This group comprises individuals from many backgrounds, including design and consulting engineering, the construction industry, equipment manufacturing, pipeline owners, and academia.

Principal Authors

Jon Robison, P.E., *MOP Task Committee Chair*—GeoEngineers, Inc.
Jeff Scholl, P.E., *MOP Task Committee Vice Chair*—JD Hair and Associates
Andrew Sparks, P.E., *MOP Task Committee Secretary*—Laney Directional Drilling
Joachim Engelhardt—Herrenknecht, AG
Andrew Finney, P.E.—Jacobs
Norm Joyal, P.E.—McMillen Jacobs Associates
Chris Lamont, P.Eng.—Associated Engineering Alberta Ltd.
Michelle Macauley, P.E.—Macauley Expert Services
Mary Neher, P.E.—Bennett Trenchless Engineers
Matt Smith—Michels
Nic Strater, P.G.—Brierley Associates
Webb Winston, P.E.—Williams

Other individuals who served and contributed

Abner Chen, Ph.D., P.E.—Purdue University Northwest
Glenn Duyvestyn, Ph.D., P.E., P.Eng.—Orsted
Michael Martens, P.Eng.—TC Energy
Rob Tumbleson—ECI Contracting

Jwala Sharma, Ph.D., P.E.—Michels
Angela Williamson, P.E.—Kinder Morgan

A thank you to the members of the Blue Ribbon Review Committee who reviewed the finished document and provided their comments:

Alan Snider, P.E., *Chair*—The Trenchless Company, LLC
Samuel Ariaratnam, Ph.D., P.E.—Arizona State University
Glenn Boyce, Ph.D., P.E.—McMillen Jacobs Associates
Gerhard Lang—Herrenknecht, AG

A thank you to Julie McCullough (McMillen Jacobs Associates) for providing technical editing, and to James Chae, P.E. (Jacobs) for editorial review.

In addition to acknowledging the support provided by all those companies whose employees participated in the development of this manual, the task committee would like to specially acknowledge the significant support provided by the following companies:

GeoEngineers, Inc.
Herrenknecht, AG
McMillen Jacobs Associates
Michels
Williams

ASCE UESI Committee on Trenchless Installation of Pipelines (TIPS):

Jon Robison, P.E., *Chair*
Jeff Boschert, P.E., *ExCom Liaison*

CHAPTER 1
INTRODUCTION AND HISTORY

1.1 INTRODUCTION

Pipelines of various types, sizes, and purposes are routinely installed below the ground surface in every community around the world. The most common method of pipeline installation is open cut, also known as trenching. However, open-cut installation results in obvious and sometimes significant unwanted ground disturbance. Consequently, numerous trenchless methods of pipeline installation have been developed to reduce surface impacts and public inconvenience (Table 1-1).

Direct steerable pipe thrusting (DSPT) was developed to fill a need for a pipeline installation in ground conditions challenging to other trenchless methods. Direct steerable pipe thrusting is a relatively recent innovation in the trenchless industry combining some of the characteristics of horizontal directional drilling (HDD) and conventional microtunneling. Direct Pipe, developed by Herrenknecht AG (Herrenknecht) in Germany, was the first commercially available system to fall into the DSPT category of trenchless technology. As described in Section 1.2, the first Direct Pipe installation was completed in Germany in 2007; as of 2022, more than 200 installations have been completed worldwide. This Manual of Practice (MOP) reflects the current state of industry development.

1.1.1 Definition of Direct Steerable Pipe Thrusting

Direct steerable pipe thrusting is a near-surface launched, thrusted microtunnel method of pipeline installation that can install a steel pipe in a single pass along both horizontal and vertical curves.

Table 1-1. Common Sizes, Lengths, and Capabilities of Various Trenchless Methods.

Method	Size, OD	Length	Groundwater	Grade control
Pipe ramming[a]	8–96 in. (205–2,440 mm)	250 ft (75 m)	Both[b]	No[c]
Horizontal auger boring	12–72 in. (305–1,830 mm)	300 ft (90 m)	Above	No
Guided boring	4–48 in. (100–1,220 mm)	300 ft (90 m)	Both	Yes
Microtunneling[d]	30–96 in. (760–2,440 mm)	1,500 ft (460 m)	Both	Yes
Shielded pipe jacking	48–120 in. (1,220–3,050 mm)	1,500 ft (460 m)	Above	Yes
Direct steerable pipe thrusting	36–56 in. (915–1,420 mm)	4,000 ft (1,220 m)	Both	Yes
Horizontal directional drilling	4–42 in. (100–1,065 mm)	6,500 ft (1,980 m)	Both	No
Utility tunneling	48 in. or larger (1,220 mm or larger)	Any length	Above	Yes

Note: These are common parameters and do not represent absolute minimum and maximum, nor do they consider additional ground improvement methods or other risk mitigation methods. These values are subject to change as technology evolves. The table is based on a similar table found in ASCE MOP 133 (ASCE 2017).

[a]Length decreases as diameter increases.

[b]Below the groundwater may be possible, but it is limited by ground type or depth below groundwater.

[c]Limited grade control may be achievable through steering heads or hybrid methods.

[d]Length increases as diameter increases.

Direct steerable pipe thrusting can be characterized as follows:

1. Soil or rock in the heading of the pipe is removed through slurry microtunneling methods.
2. Clamping device on the exterior (perimeter) of the pipe is used in combination with a thrusting machine to hold and push the installation pipe along its heading.
3. Excavated materials are removed by a closed slurry system located within the installation pipe.
4. Installation pipe is steel and is prefabricated (typically on site) in one or more sections.
5. Pipe is installed in compression by using a thrusting machine.

1.1.2 Purpose and Organization of the Manual of Practice

The purpose of this manual of practice is to provide general information and design and construction guidance for the DSPT installation method to engineers, owners, contractors, equipment manufacturers, and other project and industry stakeholders. This manual is not a standard, nor do the authors believe it defines the standard of practice; rather it is intended to provide information on the DSPT method and serve as a contribution to current DSPT engineering practice. Accordingly, the authors have chosen to give recommendations, where appropriate, but have not elected to phrase the text in terms of requirements; this is in keeping with longstanding ASCE MOP tradition. This manual is divided into the following chapters:

- Chapter 1: Introduction and History
- Chapter 2: Method
- Chapter 3: Planning
- Chapter 4: Detailed Site Investigations and Geotechnical Engineering
- Chapter 5: Engineering Design
- Chapter 6: Steel Pipe Design
- Chapter 7: Contract Documents
- Chapter 8: Construction Considerations
- Glossary

For information on microtunneling and HDD methods, the reader may refer to the following:

Microtunneling: ASCE Standard 36-15 (ASCE 2015) and
HDD: Pipeline Design for Installation by Horizontal Directional Drilling, MOP 108 (ASCE 2014) or *NASTT's Horizontal Directional Drilling (HDD) Good Practices Guidelines* (NASTT 2017).

1.2 HISTORY OF DIRECT STEERABLE PIPE THRUSTING

The history of DSPT (Direct Pipe) begins with two entirely different underground installation methods: HDD and microtunneling.

1.2.1 Horizontal Directional Drilling

Since its inception in the 1970s, HDD has been one of the preferred trenchless installation methods when crossing rivers and other natural and man-made obstacles. HDD is a very effective and economic trenchless installation method, originally developed by Martin Cherrington. HDD technology and the capabilities of HDD contractors have improved to the point that pipelines up to 60 in. (1,525 mm) in diameter and crossings over a mile in length have been completed.

Installation of pipelines by HDD is a multistep process. As shown in Figure 1-1, first, a small-diameter pilot hole is drilled. This initial bore is then enlarged to the final diameter required to install the pipeline in one or more steps of reaming. During drilling, the hole is uncased and must be kept open by using a soil-pressure-balancing drilling fluid.

One advantage of HDD is that almost all the required equipment and components are located aboveground during drilling. The underground components include drill rods, bits, reamers, and swabbing tools. Although this is advantageous because shafts are not required, and some of the more

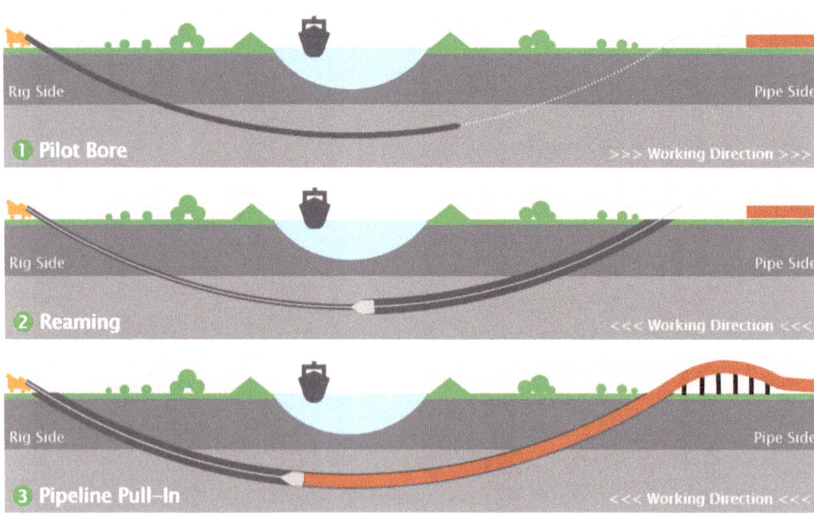

Figure 1-1. Horizontal directional drilling (HDD) process: (1) pilot bore, (2) reaming, (3) pipeline pull-in.
Source: Courtesy of Herrenknecht.

complicated and therefore failure-prone equipment is easily accessible for repairs and adjustments during construction, there are some drawbacks. Principal among these is that HDD requires the excavated cuttings to be transported (or *returned*) to the surface through the annulus of the uncased borehole, between the drill rod and the adjacent formation. Consequently, drilling fluid has two primary purposes: (1) it stabilizes the borehole; and (2) it transports the cuttings to the surface. In challenging ground conditions (e.g., where the ground conditions are very soft or very loose), or in areas of low overburden (e.g., in the beginning of a borehole), the drilling fluid pressure required to transport cutting-laden drilling fluid can exceed the confining capacity of the surrounding geology. This, in turn, can lead to hydraulic fracturing of the ground and eventual surface release of the drilling fluid. An inadvertent drilling fluid surface release is commonly referred to as a frac-out. Although a drilling fluid surface release is usually nonhazardous, it can cause environmental distress, public concern, and preferential pathways for seepage.

By its nature, HDD requires workspace on both ends of the alignment. Typically, the workspaces are referred to as a rig side and a pipe side, and the drill rig is used to pull the installation pipe from the pipe side to the rig side. To reduce the risk of the borehole collapsing or other problems being encountered during the pipe pullback operation, the pipe is, in general, fabricated into as few sections as possible prior to pullback so that the installation can be completed as quickly as possible. In an ideal situation, this means that a pipe side work area is at least as long as the total assembled pipe string. If pipe side workspace is limited, HDD installations may be completed with pauses during pullback to accommodate intermediate welds or fusions, but a significant length of work area roughly parallel to the crossing is still typically required.

The accuracy of HDD depends on the project geology, steering tools, tooling, operator skill, and design geometry. It is usually expected that a reasonable accuracy for HDD will be installed within approximately 10 ft (3 m) of design line and grade.

1.2.2 Microtunneling

Microtunneling was developed in the 1970s in Japan by combining shielded tunneling methods and pipe jacking and introducing remote control technologies. It was refined in Germany in the early 1980s and was introduced to North America in 1984.

Microtunneling is a remote-controlled, guided pipe jacking system that provides continuous pressure to the face of the excavation to balance earth and groundwater pressures. Typical microtunnel diameters range from 24 to 96 in. (0.6 to 2.4 m). Individual drive lengths vary and depend on the project geology, topography, pipe diameter, pipe materials, tracking equipment capabilities and accuracy, and cutter tool wear (including underground tool replacement capabilities such as hyperbaric interventions), but range from about 500 to 1,500 ft (150 to 500 m) for pipe diameters up to 60 in. (1.5 m). The majority of microtunnel installations are straight; however, curved drives are becoming increasingly common.

As shown in Figure 1-2, microtunneling involves jacking pipe one joint at a time from an excavated shaft with a limited footprint. Microtunneling offers several advantages compared to HDD. First, the installation pipe is installed at the same time the tunnel is excavated, thus stabilizing the ground and reducing the risk of surface settlement and heave. Furthermore, the excavated ground (spoils) is returned by a closed slurry circuit consisting of a feed and return line located inside the installation pipe. A second, independent fluid line provides lubrication fluid to the annulus between the excavated ground and the jacked pipe. These fluid systems work independently, so the spoils can be removed without adding fluid pressure to the annulus, thus reducing the risk of hydraulic fractures and frac-outs. Microtunneling involves the direct-jacking of the pipe, referred to as a one-pass process. The jacked pipe can also be a casing, with the carrier pipe installed, referred to as a two-pass process.

Because the slurry and lubrication fluid pressure counterbalance the groundwater pressure, the risk of hydraulic fracture is significantly reduced compared to HDD. This allows for shallow (relative to HDD) alignments and gives microtunneling a wide spectrum of geological conditions to which it can be applied. In addition, microtunneling is

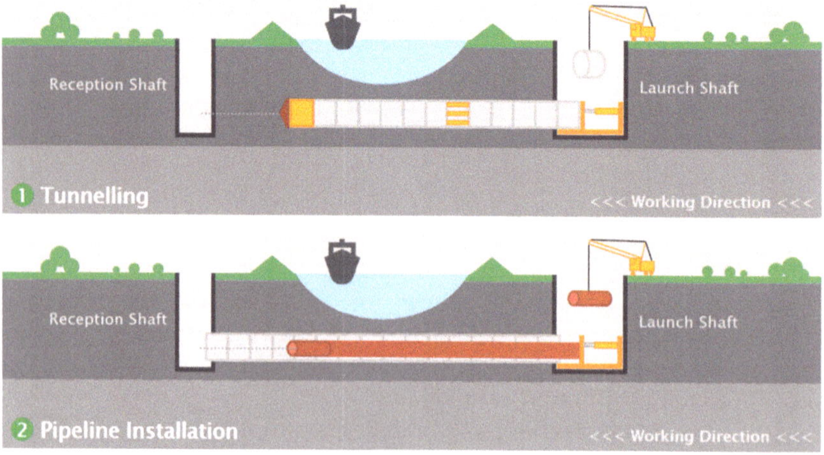

Figure 1-2. Microtunneling: (1) tunneling, direct-jacked one-pass installation; (2) pipeline installation, two-pass installation.
Source: Courtesy of Herrenknecht.

constructed or operated from one side of the tunnel, with only a small reception area required on the exit side to retrieve the microtunnel boring machine (MTBM).

Microtunneling can also provide much better installation accuracy than HDD because the guidance system technologies are more precise. This gives microtunneling the ability to maintain a consistent line and grade, making it well suited for the installation of gravity pipelines.

1.2.3 Horizontal Directional Drilling and Microtunneling in the Oil and Gas Industry

Despite its benefits, microtunneling has not been widely used in the oil and gas industry. This is mainly because oil and gas pipelines are typically high-pressure welded steel pipes. Oil and gas pipeline installations require thorough testing and integrity verification. The entire industry has developed procedures and sequential operational steps involving prefabrication and testing of relatively long pipe strings prior to installation, which do not lend themselves to the direct-jacked installation of a single pipe section at a time, as in typical microtunneling.

Where microtunneling is used for oil and gas pressure pipeline applications, a casing pipe is typically installed, and in a second stage, the final product pipeline is installed within the casing. This increases the installation cost and eliminates the one-step process of microtunneling. As a result, it is not typically economically feasible to use microtunneling for trenchless pipeline installations for the oil and gas industry.

1.2.4 Pipe Pushers and Thrusters

To aid pullback operations of pipelines on HDD projects, manufacturers and contractors developed equipment that would simultaneously push the pipeline into the bored hole while pulling the pipeline with the drill rig from the other end of the borehole. The most basic, and probably most common, units used a set of friction rollers or wheels to convey the pipeline. Later, more advanced units used tracks for additional grip and push force. All these had one thing in common: they conveyed the pipeline *continuously* because speed is a major factor in pipeline installation. However, these types of conveyor-style units are limited to approximately 100,000 to 250,000 lb (50 to 115 metric t) of push force. Over time, they became known as *pipe pushers*. Two types of pipe pushers are shown in Figures 1-3 and 1-4, wheel and track based. Pipe pushers are still used within the pipeline industry, especially for offshore applications.

In 1994, Martin Cherrington developed what has become known as a *pipe thruster*. It consisted of an A-frame guiding the pipeline and a mechanical collar that would attach to the pipeline. Both these would

Figure 1-3 (left). Wheel-based pipe pusher.
Source: Courtesy of Herrenknecht.

Figure 1-4 (right). Track-based pipe pusher.
Source: Courtesy of Herrenknecht.

connect through a number of sheaves and cables to a tractor on one side and a deadman anchor on the other. As the tractor pulled, the collar would tighten around the pipeline and, at the same time, pull the pipeline toward and into the borehole. The fundamental difference between a pipe thruster and a pipe pusher is that, whereas a pipe pusher provides continuous pipe movement, a pipe thruster has a limited stroke, and requires regripping of the pipe once the stroke limit is reached. Despite this limitation, pipe thrusters are typically more powerful than pipe pushers.

In 2006, German manufacturer Herrenknecht introduced its version of the pipe thruster, which features a clamping unit that hydraulically grips the pipe and then pushes it forward by means of hydraulic cylinders. A photograph of a Herrenknecht pipe thruster is shown in Figure 1-5.

The largest of the three Herrenknecht pipe thrusters can deliver up to 1.65 million lb (750 t) of thrust in both directions (push and pull) and handle pipelines up to 60 in. (1,525 mm) in diameter. This made it not only a useful assist tool for the installation of pipelines by HDD, but also a rescue tool for pipelines that became stuck in the borehole during pullback. Table 1-2 summarizes the capacity and other statistics of available Herrenknecht pipe thrusters.

In 2011, Prime Drilling introduced a conveying and clamping style pipe thruster (named the Prime Pipe Pusher) that delivers up to 2.2 million lb (1,000 t). Table 1-3 summarizes the capacity and other statistics of available Prime pipe thrusting equipment. An example of a Prime thrust unit is shown in Figure 1-6.

1.2.5 Horizontal Directional Boring

As early as the late 1970s, Martin Cherrington combined key features from HDD, pipe jacking, and oilfield drilling methods to develop and

Figure 1-5. Pipe thruster.
Source: Courtesy of Herrenknecht.

patent horizontal directional boring (HDB), an installation method for large-diameter pipelines in soft formations over long distances. This was the very first one-step installation technique using a thrust-forward device.

The original HDB system consisted of a thrusting unit, an independently rotating drill head with fluid nozzles, a cuttings vacuum suction port, and a directionally controlled and articulated steering head. The drill head was attached to the product pipe (or casing). Unlike HDD, the drilling mud

Table 1-2. Herrenknecht Pipe Thrusters.

Pipe thruster	HK 300 PT	HK 500 PT	HK 750 PT
Push and pull force	661,000 lb (300 t)	1,102,000 lb (500 t)	1,653,000 lb (750 t)
Stroke	13.12 ft (4 m)	16.4 ft (5 m)	16.4 ft (5 m)
Pipeline diameter	8–36 in. (205–915 mm)	8–48 in. (205–1,220 mm)	8–60 in. (205–1,525 mm)
Approximate weight	81,600 lb (37 t)	99,200 lb (45 t)	1,873,400 lb (85 t)
Operating angle[a]	0–15 degrees	0–15 degrees	0–15 degrees

Source: Courtesy of Herrenknecht.

[a]Although the pipe thruster is mounted horizontally level, the operating angle can be continuously adjusted within this range. For additional angles, the pipe thruster must be mounted declined.

Table 1-3. Prime Drilling Pipe Pushers.

Prime thrust unit designation	PPP400	PPP500	PPP1000
Push and pull force	881,000 lb (400 t)	1,102,000 lb (500 t)	2,204,000 lb (1,000 t)
Stroke[a]	Variable	Variable	Variable
Pipeline diameter	12–60 in. (305–1,525 mm)	12–60 in. (305–1,525 mm)	12–60 in. (305–1,525 mm)
Approximate weight	88,200 lb (40 t)	99,200 lb (45 t)	121,250 lb (55 t)
Operating angle[b]	0–45 degrees	0–45 degrees	0–45 degrees

Source: Courtesy of Prime Drilling.

[a]Pusher operates on a rack and pinion drive.

[b]Pusher must be mounted on a decline to achieve the desired operating angle.

was extracted from the annulus through the vacuum port into the cutting head and pumped internally to the surface through a dedicated return line, thus reducing the risk of hydraulic fracture normally associated with pumping the returns through the open borehole annulus. A second dedicated internal line supplied the drilling mud to the cutting head, and

Figure 1-6. Prime pipe pusher.
Source: Courtesy of Prime Drilling.

a third dedicated internal line supplied lubrication mud to the bored hole annulus near the leading end of the bottom-hole-assembly. Hydraulic lines from the surface were fed through the thrusted pipeline and supplied power for rotating the cutting wheel and operating the steering cylinders of the articulated head, and a pump for collecting and returning the cuttings to the surface (via a fourth dedicated line). Additional pressure sensors helped maintain the balance between hydrostatic annular pressure and formation pressures. Stabilizers and enhanced directional surveying instrumentation to control the drill rate and to perform needed alignment corrections completed this ambitious but ingenious setup. It did not feature a crushing chamber like microtunneling systems would have some years later, so any cobbles or boulders would either need to be broken by the cutting wheel or pushed out of the bore path. Nevertheless, it was the first single-step trenchless installation method for pipelines incorporating key elements from what would become DSPT.

Two successful HDB applications were completed. The first was on Greens Bayou, east of Houston, in Proctor and White (1977), where 750 ft (230 m) of a 30 in. diameter casing (765 mm) for a 24 in. (610 mm) diameter crude oil pipeline was installed for the Colonial Pipeline. The second was to place a 40 in. (1,000 mm) diameter gas pipeline 2,100 ft (640 m) across the Houston Ship Channel for the Colonial Pipeline in 1978. Both crossings were through clay and confirmed HDB as a viable alternative expanding the limits of trenchless technology at the time. After these two successful projects, HDB was put back on the shelf, because increasingly efficient, multistep HDD techniques were being introduced. However, HDB laid the foundation for what would later be called *extended reach drilling*, now used extensively in the oilfield industry to drill wells horizontally 10 to 12 mi (15 to 20 km) or greater.

1.2.6 Easy Pipe: The Beginning of Direct Steerable Pipe Thrusting

With the introduction of the pipe thruster, it became evident that it should be possible to use a microtunnel boring machine (MTBM) for traditional HDD projects, while overcoming some of HDD's shortcomings. Inspired by this idea, a trenchless method known as Easy Pipe (patented by German contractor Meyer & John) was invented in 2006. It consisted of a jacking process whereby bolted steel tubes (replacing the steel casing or other jacking pipes in traditional microtunneling) were pushed toward the reception or exit point, connected to the installation pipe, and then pulled back together toward the launch or entry point (Figure 1-7).

From 2008 until 2011, there were eight 56 in. diameter crossings completed with Easy Pipe for the OPAL pipeline in Germany. These crossings ranged from 150 to 395 ft (45 to 120 m) in length, and the total length installed on the project was 1,700 ft (520 m). The maximum depth

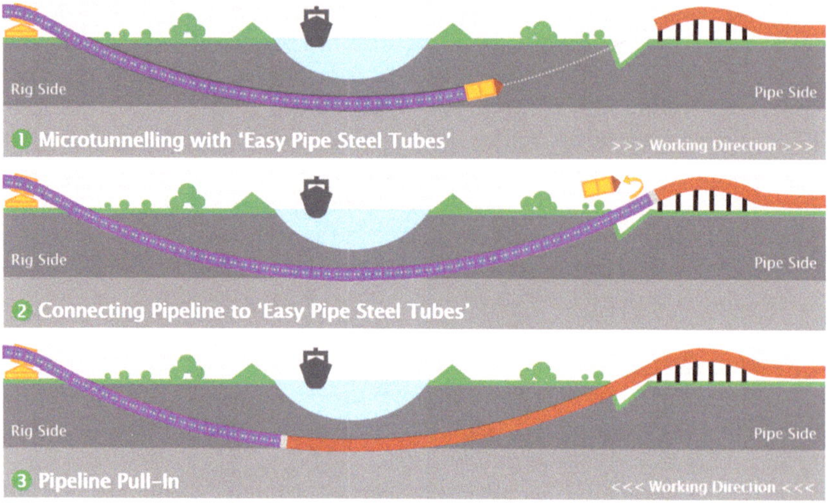

Figure 1-7. Easy pipe: (1) tunneling with easy pipe steel tubes, (2) connecting pipeline to easy pipe steel tubes, (3) pipeline pull-in.
Source: Courtesy of Herrenknecht.

of these crossings was around 16.5 ft (5 m) in silts, sands, gravels, and cobbles. Two of the crossings were under rivers and six passed beneath railroad tracks.

However, Easy Pipe is a two-step process and requires the use of specialized steel tubes, which are reusable but fixed to one diameter. In the years to follow, the existing easy pipe equipment was eventually converted to Direct Pipe, allowing welded pipe and multiple diameters. In 2008, Meyer & John patented another trenchless installation method called *Easy Long*, a derivative of Easy Pipe (Section 1.2.8).

1.2.7 Direct Pipe: Herrenknecht Developments

Along with their pipe thruster (Section 1.2.4), Herrenknecht introduced the present-day concept of Direct Pipe in 2006. By combining microtunneling technology with the curved profile of HDD, a single-pass trenchless installation method for steel pipelines was created, shown in Figure 1-8.

The Direct Pipe method includes a prewelded and tested pipeline (thus reducing the number of golden welds) and a DP-MTBM that is attached to the front of the pipeline. The pipe thruster operates as a thrust unit from a launch pit, clamping the pipeline on the outside and pushing the DP-MTBM and the pipeline into the ground. The bored hole is excavated by the DP-MTBM and, like all slurry-based machines, the tunnel or bored

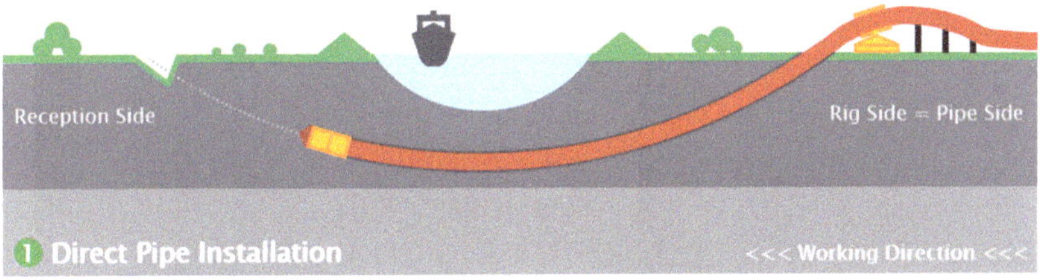

Figure 1-8. Direct Pipe: direct pipeline installation.
Source: Courtesy of Herrenknecht.

hole face is supported by the slurry. The Direct Pipe method allows for straight or curved boreholes (in both the vertical and the horizontal planes).

Compared to HDD, Direct Pipe typically requires lower downhole fluid pressures and usually presents less risk of frac-out. Depending on ground conditions, this allows for reduced installation depths. In many cases, it is also more accurate than HDD, and with continuous applied face pressure, it presents a reduced risk of bored hole collapse. At the same time, as shown in Figures 1-9 and 1-10, it provides attributes similar to HDD, such as surface or near-surface launching at an angle.

Compared to microtunneling, Direct Pipe allows surface launch and retrieval of the DP-MTBM, eliminating the need for deep shafts. It also provides the ability to retract the DP-MTBM, reducing the risk presented by obstructions and the need for rescue shafts to access the equipment.

Direct Pipe is often viewed as being similar to HDD because of the curved profile, the pipeline handling, and the welding process. However,

Figure 1-9. Typical construction layout of Direct Pipe.
Source: Courtesy of Herrenknecht.

Figure 1-10. Pipeline preparation and stringing.
Source: Courtesy of Herrenknecht.

because it uses microtunneling technology to excavate the ground, it is more closely related to traditional microtunneling. Only the pipe thruster has been borrowed from its initial HDD application. To date, the majority of the Direct Pipe applications have been within the domain of HDD's primary market. This is evident by the early installers of Direct Pipe, who were predominantly HDD contractors familiar with oil and gas projects and welded steel pipeline handling.

The very first Direct Pipe pilot project was near the city of Worms in Germany in 2007, crossing the Rhine River in cobbles, gravels, and sands. A 48 in. (1,200 mm) diameter steel casing, more than 1,500 ft (460 m) long, was installed with Direct Pipe, to serve as a casing for a water pipeline and several ducts for power transmission and telecommunication cables. The depth of the ground cover under the river was approximately 33 ft (10 m), and 10 ft (3 m) below the riverbanks. The water depth of the river was approximately 16 ft (5 m).

The Rhine River pilot project was completed with a bare, uncoated casing pipe. To demonstrate that Direct Pipe could be used with coated pipelines, Herrenknecht completed tests on steel pipe using different coatings, including polyethylene (PE) and with glass-fiber-reinforced plastic coatings (GRP on a PE sublayer). The testing revealed that the pipe thruster would not harm the coatings, even at full clamping force.

With the coating tests completed, the second Direct Pipe project was completed in 2009. This involved a river outfall with a 48 in. (1,200 mm) diameter PE-coated steel casing, which was designed to contain a 36 in. (915 mm) diameter GRP-coated product pipe for an underground natural gas storage facility inside a salt cavern. The crossing was 920 ft (280 m) long, and the pipe was fabricated in 120 ft (36 m) sections. In this case, Direct Pipe was considered the preferred trenchless method because of the potential for unforeseen obstructions such as man-made shoring timbers and buried stone structures, and environmental concerns associated with the risk of drilling fluid surface release.

The successful completion of these projects set the stage for the next phase of Direct Pipe utilization: direct, single-pass installation of coated product pipelines for the oil and gas industry. In 2009, after extensively testing the integrity of polypropylene (PP)-coated 48 in. diameter pressurized steel pipelines, a Dutch gas supplier completed several Direct Pipe installations during the expansion of its distribution network. By the end of 2010, five more Direct Pipe crossings on three different projects had been completed. These crossings ranged in length from 1,180 ft (360 m) to 1,770 ft (540 m) and crossed below rivers, railroad tracks, and archaeological sites up to a 100 ft (30 m) depth. Over the next several years, additional tests demonstrated that Direct Pipe could be successfully used with other coatings such as fusion-bonded epoxy (FBE) and concrete coating.

The first Direct Pipe projects in the United States were completed during 2010. The first crossing involved installation of a 30 in. (750 mm) diameter, FBE-coated gas transmission pipeline in Florida. By early 2011, three crossings had been completed in Florida that ranged in length between 390 and 740 ft (120 and 225 m). The geotechnical conditions for these crossings consisted of sand, clay, and coral. The next crossing was a 42 in. (1,070 mm) diameter casing installation in upstate New York of more than 425 ft (130 m) in gravels and cobbles.

In late 2011, Direct Pipe was used in another application, installing a conductor casing for an HDD crossing. A project on the Hudson River in the New York–New Jersey area encountered near-surface ground conditions including fill, wood, sand, cobbles, and boulders that were deemed unsuitable for HDD. In this case, a 48 in. (1,200 mm) diameter steel conductor casing was installed with Direct Pipe over 280 ft (85 m) through these materials. After the casing was installed, the DP-MTBM was retracted through the casing, enabling HDD to complete the project.

In 2011, the first truly long-distance installation of a 48 in. diameter (1,200 mm) steel pipe with PP coating took place for a gas pipeline. A record setting 4,595 ft (1,400 m) crossing near Lochem, Netherlands, used two 500 t pipe thrusters in series.

In 2014, a 2,445 ft (745 m) crossing with an FBE-coated, 48 in. (1,200 mm) diameter gas pipeline was successfully completed underneath the Rio Grande in Texas, where the entry point was on the territory of the United States and the exit point was located in Mexico. This crossing was completed with one single pipe string without additional welding (hence, no golden welds) after launching the DP-MTBM.

Also in 2014, the first two Direct Pipe projects were completed in Canada, including a 42 in. (1,200 mm) diameter gas pipeline in Bonnyville, Alberta, and a 48 in. (1,200 mm) diameter casing in Calgary, Alberta. Worldwide, more than 20 mi (30 km) of pipeline from 36 to 56 in. (750 to 1,400 mm) in diameter had been installed by Direct Pipe by the end of 2015.

In 2015, two Direct Pipe installations were completed below levees within the jurisdiction of the US Army Corps of Engineers (USACE). The USACE is sensitive about the potential for elevated annular drilling fluid pressures associated with HDD, which can damage levees and their foundations. Direct Pipe was selected for these installations because, compared to HDD, it requires much lower downhole pressures.

One of these two projects was in Port Arthur, Texas, crossing the Sabine Neches Waterway with a 48 in. (1,200 mm) diameter steel casing for a 26 in. (650 mm) diameter high-density polyethylene water line. At the time, it also set the North American distance record for a Direct Pipe installation with 3,445 ft (1,050 m) and contained two horizontal curves. This record crossing was succeeded a few weeks later by a Direct Pipe installation of a 48 in. diameter casing for a 16 in. (405 mm) diameter gas pipeline in Freeport, Texas, crossing the Dow Barge Canal with a distance of 4,035 ft (1,230 m). The DP-MTBMs on both projects were equipped with transducers recording the outside pressure of the lubrication fluid in the annulus between the bored hole and the DP-MTBM to document conformance with USACE permit requirements.

In 2017, a 48 in. (1,200 mm) diameter Direct Pipe crossing of Highway 16 near Edson, Alberta, was completed at a length of 710 ft (215 m). The project was under the jurisdiction of the National Energy Board (NEB) of Canada. After an evaluation of the Direct Pipe installation method, NEB officially recognized and approved Direct Pipe for trenchless pipeline construction in Canada.

By early 2018, approximately 30 mi (50 km) of Direct Pipe installations had been completed in various ground conditions ranging from sands, silts, and clays to near zero blow count soft soil, different types of rock, and various amounts of gravels, cobbles, and boulders. Although some of the completed crossings had difficulties, nearly all were successful. Only two Direct Pipe crossings were not completed because of unfavorable subsurface conditions; both these projects were in ground with high concentrations of boulders. These failures demonstrated that Direct Pipe, like all trenchless methods, is not always feasible in every ground condition.

In addition to land-to-land crossings, in recent years, interest has grown in the application of the Direct Pipe method to pipeline shore approaches and outfalls. Some of these applications include sea outfalls from water desalination plants and shore approaches for offshore windfarms for renewable energy, ocean lay pipelines, or offshore loading structures.

Direct Pipe construction is largely independent of weather and has the advantage of being a land-based operation that only requires offshore support to recover the DP-MTBM. The environmental impact of Direct Pipe is, in general, relatively low because the risk of disturbing sea life and destroying marine habitats (compared to dredging) or from releasing drilling fluid into the water is low compared to HDD because of the relatively low fluid pressures, small overcut, and accurate control of slurry volume.

In 2017, 10 years after the very first pilot project, Direct Pipe was used for the first time on a true offshore project in Scotland, United Kingdom. A Norwegian contractor chose Direct Pipe to bring in the export cables from an offshore windfarm through two 48 in. diameter steel casings, each about 1,440 ft (440 m) long. The DP-MTBM was recovered by divers.

Later in the same year, a Canadian pipeline company, already experienced with Direct Pipe, contracted with a United States-based contractor to install almost 5,000 ft (1,500 m) of 42 in. (1,050 mm) diameter gas pipeline with FBE coating, near Port Isabel, Texas, claiming a new long-distance record. The Direct Pipe shore approach crossing exited the sea floor into the Gulf of Mexico in about 20 ft (6 m) of water and the DP-MTBM was lifted onto a barge for removal from the pipeline.

In the meantime, more shore approaches have been completed successfully. Most notable are two Direct Pipe installations in New Zealand in 2018 and 2020. Both were 48 in. diameter wastewater replacement outfalls—the first one was over 6,330 ft (1,930 m) at Army Bay near Auckland, whereas the other went over 6,630 ft (2,020 m) near Snells Algies into Martins Bay (also on the North Island).

The success of Direct Pipe installation over a wide variety of site and subsurface conditions, in addition to developments for offshore applications and experience by contractors and owners, has made these more recent achievements possible.

1.2.8 Direct Steerable Pipe Thrusting Variations

Since the inception of DSPT, other similar installation methods were developed. Most notable is the joint effort of two German equipment manufacturers, MTS and Prime Drilling, which launched their System2 in 2013, for which Prime Drilling supplied the pipe pusher and MTS provided the microtunneling unit. Its first project was crossing the Moskva Canal near Khimki in the greater Moscow area with a 48 in. (1,200 mm) diameter casing for a gas pipeline to St. Petersburg. The 607 ft (185 m) crossing was completed in 40 ft (12 m) sections of slightly prebent pipes (to reduce the borehole distance by allowing for a tighter bend radius).

In 2008, another patented method named Easy Long mimicked and expanded on Easy Pipe (Section 1.2.6). It is also a jacking process but is

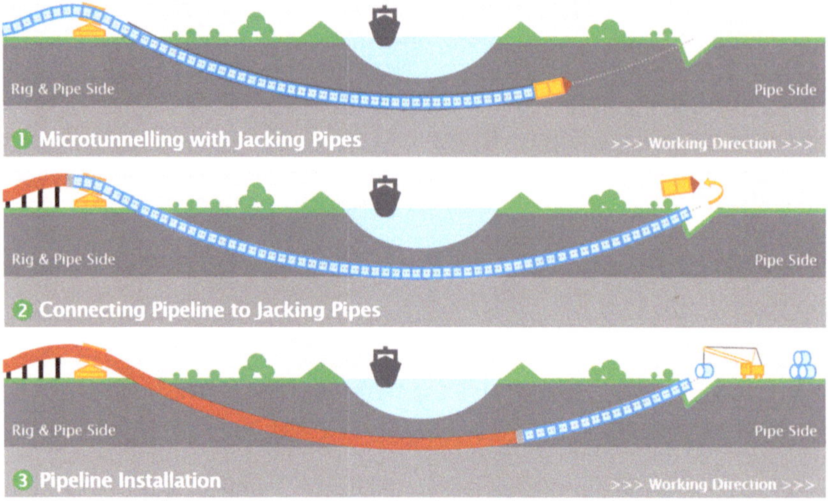

Figure 1-11. Easy long: (1) microtunneling with jacking pipes, (2) connecting pipeline to jacking pipes, (3) pipeline installation.
Source: Courtesy of Herrenknecht.

different from Easy Pipe, which uses tightly connected steel tubes for push and pull operation. Easy Long uses reinforced concrete pipe jacking pipes (e.g., from microtunneling) or similar, widely available jacking pipes, which are pushed toward the reception point, as shown in Figure 1-11. Then, the pipeline is pushed from the launch point behind the jacking pipes until it (and all the jacking pipes) arrive at the receiving side. It expands further by allowing the system to push the pipeline into the tunnel from the reception side (instead of launch side) or to pull the pipeline from the reception side to the launch side (like Easy Pipe).

1.3 CONCLUSIONS

Direct steerable pipe thrusting has proven over the years that it is a valuable tool for installing underground pipelines by a single-pass trenchless pipeline method. It has the potential to overcome some of the limitations of both HDD and microtunneling in many cases. However, DSPT technology is not the installation solution for every potential project; it too has its limitations.

Direct steerable pipe thrusting is an advanced technology requiring detailed knowledge about the method, its technical aspects, the process, and the methodology. One should not approach DSPT with HDD principles, at least not exclusively. Direct steerable pipe thrusting technology is more akin to microtunneling technology. However, not all principles from microtunneling are valid for DSPT.

History has shown that new trenchless technologies, adaptations, and applications require new approaches to design and construction. Everyone is encouraged to read this MOP and to become familiar with the current state of practice in the industry.

REFERENCES

ASCE. 2014. *Pipeline design for installation by horizontal directional drilling*, 2nd ed. MOP 108. Reston, VA: ASCE.

ASCE. 2015. *Standard design and construction guidelines for microtunneling*. ASCE 36-15. Reston, VA: ASCE.

ASCE. 2017. *Pilot tube and other guided boring methods*, MOP 133. Reston, VA: ASCE.

NASTT (North American Society for Trenchless Technology). 2017. *Horizontal directional drilling (HDD) good practices guidelines*, 4th ed. Cleveland: NASTT.

Proctor, R. V., and T. L. White. 1977. *Earth tunneling with steel supports*. Youngstown, OH: Commercial Shearing.

CHAPTER 2
METHOD

2.1 INTRODUCTION

The direct steerable pipe thrusting (DSPT) method presents a set of new options for the pipeline installation market by combining the benefits of widely accepted trenchless technologies to form a new methodology. It utilizes microtunneling technology for excavation, continuous face and wall support, and spoils removal, and it incorporates elements from horizontal directional drilling (HDD), such as prefabricated pipe string sections and near-surface-to-surface entry and exit. Unlike HDD, which typically pulls plastic or steel pipe, DSPT utilizes a pipe thruster or pusher that grips the outside of the thrust pipe to push it forward through the bored hole. This allows the thrust pipe to be installed simultaneously during the excavation of the tunnel face. The thrust pipe serves as the final product pipe or as a casing, depending on the application.

This chapter provides a summary of the DSPT process, the DSPT microtunnel boring machine (MTBM) and other equipment technologies, and some general design parameters.

2.2 INSTALLATION PROCESS

Direct steerable pipe thrusting incorporates the sequential process of a typical HDD installation including site preparation, prewelding (long) sections of thrust pipes, integrity testing and certification of joints within these long sections, application of joint coating, tunneling, and pipe installation. Even the necessity for intermediate or golden welds, nondestructive testing of the welds, and final field-joint coating remains unchanged in most cases.

Direct steerable pipe thrusting differentiates itself from traditional multistep processes such as HDD by combining the process of tunneling and simultaneously installing the thrust pipe in a one-pass installation. The key elements for this process are the MTBM (Component 1) and the pipe thruster or pusher (Component 2), as illustrated in Figure 2-1.

Comparable to a traditional jacking frame used in microtunneling or pipe jacking for pushing short pipe sections of 8 to 10 ft (2.5 to 3.0 m) in length at a time, the pipe thruster or pusher provides the forward thrust in DSPT to the prefabricated thrust pipe (Section 2.3.1). However, instead of applying force to the end of the pipe (as in conventional microtunneling), the pipe thruster or pusher clamps and grips the outside of the thrust pipe, pushing it forward into the ground.

The joining of prefabricated pipe strings, as shown in Figure 2-1, requires connecting the internal umbilical lines of the MTBM (Section 2.3.6), which run through the full length of the pipe strings, and then joining the strings together with golden welds. Section welding is typically followed by nondestructive testing/nondestructive examination procedures and field-coating of the joint(s).

The tunnel is excavated by the MTBM, which is based on slurry-type microtunneling machines and cutter heads (or cutter wheels) (Section 2.3.2). The MTBM is attached to the forward end of the thrust pipe and they are pushed into the ground together.

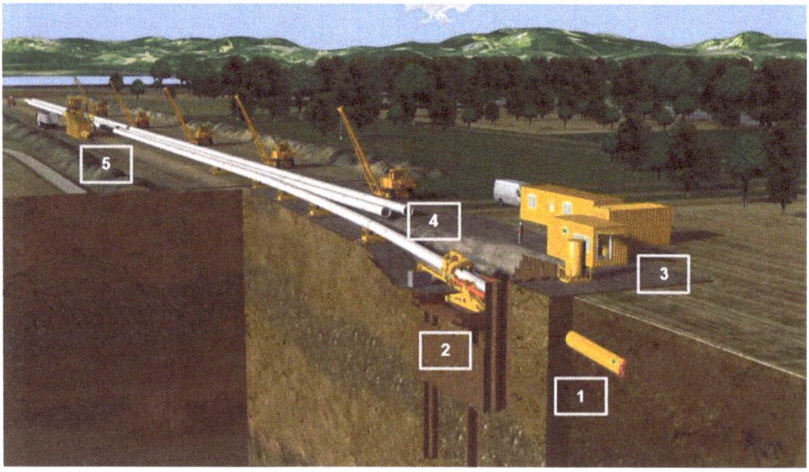

Figure 2-1. DSPT components: (1) MTBM, (2) pipe thruster, (3) control and power containers, (4) prefabricated pipe strings, (5) separation plant.
Source: Courtesy of Herrenknecht.

A rotating cutter wheel excavates the ground at the face of the tunnel. Excavated material is broken down by the cutting tools before entering the crushing chamber where it mixes with drilling fluid and passes into the slurry circuit through openings in the crushing chamber. The slurry is then pumped back through a dedicated discharge line inside the thrust pipe to the separation plant on the surface. At the separation plant, the cuttings are removed from the slurry, and the clean drilling fluid is then returned to the face of the MTBM through a dedicated feed line also located inside the thrust pipe (Section 2.3.6).

During the excavation process, the overcut or annular space (annular gap between the bored hole to the outside diameter of the MTBM or the pipeline) created by the cutter wheel gauge cut is filled with a bentonite-based lubricant. This lubricant reduces friction between the bored hole and the thrust pipe, and if the lubricant is maintained at a pressure that is slightly above ambient ground pressure, it supports the ground to maintain the integrity of the surrounding formation. The bentonite lubrication system is separate from the feed and slurry return system (Section 2.3.3). All the control and support units are located aboveground near the launch area, workspace permitting.

2.2.1 Drilling, Slurry, and Lubrication Fluids

In DSPT, the primary fluids are defined based on their function: drilling fluid, slurry (drilling fluid with cuttings), and lubrication fluid (lubricant and excavated hole support). These terms are defined in the Glossary for ease of reference and are frequently used throughout this MOP.

Bentonite is a typical component of the primary DSPT fluids. It is a naturally occurring clay mineral with thixotropic properties when mixed with water. When hydrated, it exhibits a low viscosity when in motion and becomes highly viscous (gel-like) when at rest and no energy is applied. In DSPT, bentonite plays different roles depending on the fluid's tasks.

Drilling fluid can consist of water or water with additives such as bentonite, polymers, or other modifiers to enhance its properties. If the drilling fluid is designed for a specific property, it is considered engineered drilling fluid (EDF). Drilling fluid has multiple purposes: to assist in providing face support by counterbalancing hydrostatic pressure at the face (in conjunction with the cutting wheel providing mechanical face support); to transport excavated materials or cuttings from the MTBM to the separation plant at the ground surface; and through lubricity, to reduce wear on the cutting tools, feed/return lines, and internal slurry components such as pumps.

When the drilling fluid is mixed with cuttings, it becomes drilling slurry, which is pumped out to the separation plant at the surface (Section 2.3.8). At the separation plant, the drilling slurry is cleaned for reuse to return to the MTBM face as drilling fluid.

Fluid for pipe lubrication is different from the drilling fluid and serves two primary functions: ground support and pipe lubrication. First, the lubrication fluid fills the MTBM's annular space and pipe annulus behind the machine, enhancing hole stability and giving support to the formation, provided the lubrication is kept under pressure. Second, the lubrication fluid usually has a higher viscosity than the EDF to provide the lubricity necessary to keep thrust forces low. Lubrication can be introduced through ports in the MTBM, through a lubrication ring between the MTBM and the thrust pipe, and through a valve in the launch seal. The lubrication fluid is introduced at a pressure slightly above in situ earth and groundwater pressure.

The concentration of bentonite in the lubricating fluid is higher than in the drilling fluid or EDF because a higher viscosity is required for ground support. The rheological properties and the amount of injected lubrication fluid are adjusted for the ground conditions. The MTBM's annular space and the pipe annulus should be filled completely with lubrication fluid to reduce friction between the ground and the thrust pipe to keep thrust forces low.

2.2.2 Pull-Back and Tooling Changes

One of the advantages of DSPT is that the pipe thruster can retract the MTBM together with the thrust pipe. Retraction during the installation may become necessary for several reasons, including

- Buried object, which prevents further thrusting and pipe advancement;
- Excessive thrust forces signalling that the MTBM and thrusted pipe could be stuck;
- Equipment failure or poor assembly of components; and
- Tool replacement owing to wear or adverse ground conditions.

Not all these reasons will result in a pull-back operation. Some of these difficulties can be resolved with personnel intervention, depending on local codes and regulations and the size of the MTBM.

If retraction is required during the operation and depending on ground conditions and their behavior, it may be necessary to stabilize and fill the empty void space in front of the MTBM with an EDF that can facilitate hole stability. Generally, filling with EDF takes place through the drilling fluid (feed) line. Subject to specific equipment and site conditions, there may be additional techniques to facilitate retraction such as slowly rotating the cutting wheel to reduce the risk of cutting wheel gauge cutters catching on a protrusion from the side wall when in bedrock or in cobbly or bouldery ground.

2.3 EQUIPMENT

A full spread of DSPT equipment is characterized by a number of aboveground and belowground components (Figure 2-1). The following sections provide a comprehensive overview of the key generic elements.

2.3.1 Pipe Thruster or Pusher

The primary task of the pipe thruster or pusher is to provide sufficient thrust force without damaging the thrust pipe. The pipe thruster or pusher also provides the means of retracting the entire pipe string and MTBM back to the surface when necessary. An example of a pipe pusher is shown in Figure 2-2, and an example of a pipe thruster is shown in Chapter 1, Figure 1-5.

At the time of this publication, available pipe thruster and pusher designs need to regrip or reset their clamping unit at the end of their stroke. Endless-stroke units, such as conveyor-style units, are not available for DSPT because their available forward thrust is typically insufficient.

Figure 2-2. Pipe pusher launch workspace setup.
Source: Courtesy of Prime Drilling.

2.3.2 Microtunnel Boring Machine

Direct steerable pipe thrusting uses an MTBM similar to conventional microtunneling. However, because the DSPT method installation lengths are often greater than those of conventional microtunneling, the following items should be considered when selecting the machine:

- Design of the cutting wheel and wear;
- Radial overcut;
- Power supply to the MTBM;
- Drilling fluid supply, slurry return, and lubrication;
- Positioning, steering, and communications.

The typical sections (or cans) of a DSPT MTBM are shown in Figure 2-3. The MTBM features a full-face cutting wheel at a specified tunnel diameter, a crushing chamber where cuttings are reduced in size and mixed with the EDF, becoming slurry, which is passed through a slurry discharge system to the surface for cleaning at the separation plant. The crushing chamber is immediately behind the cutting wheel in the shield and machine can or front steering joint of the MTBM.

Cutting wheels for DSPT applications are generally comparable to those used for microtunneling, although with a typically larger overcut. Typically, equipment manufacturers offer a choice of soft ground, mixed ground, or rock-cutting wheels appropriate for the anticipated ground conditions. The durability and survivability of cutting wheel tools and other mechanical components are critical for realizing a successful DSPT installation.

The other parts of the MTBM shield and machine cans typically include the main drive (bearing and gearbox) for the cutting wheel and parts of the

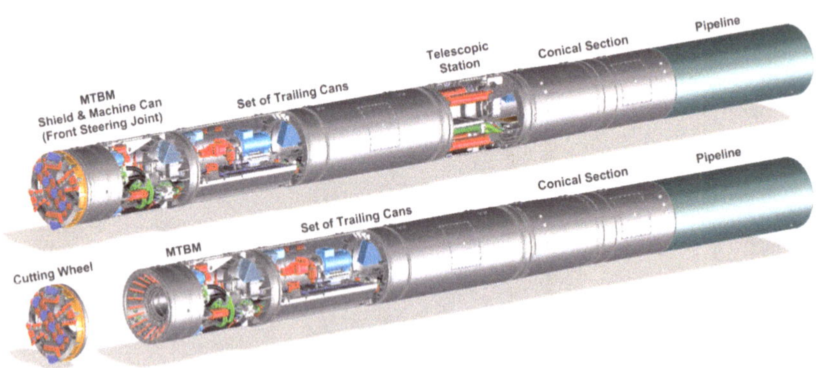

Figure 2-3. General arrangement of the MTBM (the lower image shows the MTBM without a telescopic station and with the cutting wheel removed).
Source: Courtesy of Herrenknecht.

slurry distribution components. The EDF supplied from the surface to the MTBM is directed to different locations at the front of the shield, allowing an operator to respond to varying ground conditions with the primary purpose of supporting the face and counterbalancing hydrostatic pressures. The machine can and shield includes a set of (hydraulic) steering cylinders that control horizontal and vertical direction of the MTBM to follow a predefined alignment and profile.

The MTBM shield and machine can is followed by trailing can(s) that consist of one or more articulated section(s) that house pumps, motors, electronics, communication, and other systems. For MTBMs of 36 in. (914 mm) and greater diameter, electricity is typically used to power the hydraulic systems. Smaller MTBMs are typically hydraulically powered from the surface, which limits the installation distance achieved because of hydraulic losses in the system.

Lubrication ports in one or more of the MTBM sections allow lubrication fluid to be dispensed to reduce skin friction and to stabilize the ground with a nominal pressure above ambient. The MTBM sections, whether articulated or not, are mechanically connected to one another for push and pull operations in the event a retraction becomes necessary (Section 2.2.2).

Some MTBMs contain an additional telescopic station (similar to an intermediate jacking station) serving as a forward-thrust unit. This allows the MTBM to be thrust forward without assistance from the pipe thruster at the surface. This optional feature may be used on very long crossings or in certain geotechnical conditions such as rock or other conditions where an operator prefers to have more sensitive control and feedback from the MTBM (without movement of the thrust pipe string).

In a final, usually separate, trailing section, the outside diameter of the MTBM reduces to the thrust pipe diameter. This transition or conical section is welded (or connected) to the thrust pipe. The radial gap between the excavated ground and the thrust pipe exterior wall is referred to as the pipe annulus and measures several inches on radius.

2.3.3 Pipe Lubrication

Lubricating fluid is typically introduced into the annulus surrounding the thrust pipe at two general areas: through a valve at the launch seal and through ports in the MTBM and conical section. Introduction of lubrication fluid at other locations along the thrust pipe is not common. Pipe-lubricating fluid must be introduced at a nominal pressure above ambient earth and groundwater pressure to maintain ground stability and reduce skin friction along the length of the pipeline. Figure 2-4 graphically depicts the lubrication distribution system at the MTBM.

The amount and rate at which the lubrication fluid is introduced into the annular space should be controlled, monitored, and recorded. Careful

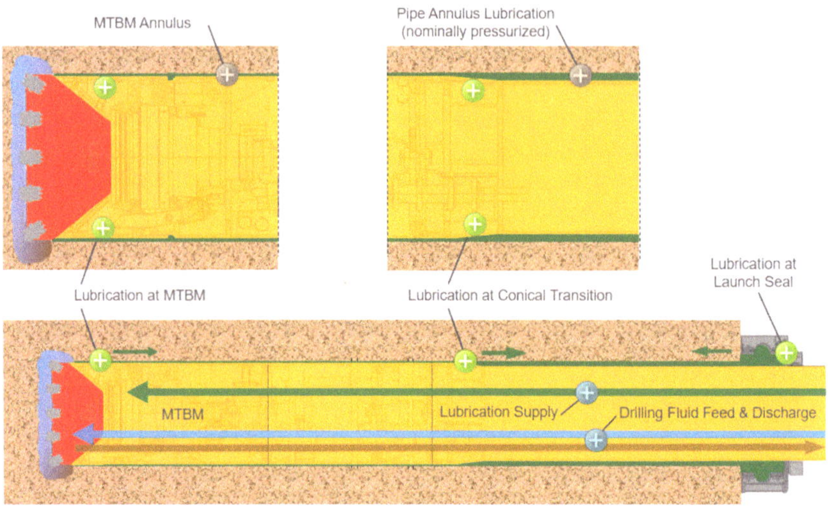

Figure 2-4. DSPT points of lubrication.
Source: Courtesy of Herrenknecht.

consideration must be given to potential losses of lubrication fluid to the surrounding ground as well as its stability over time (e.g., degradation through saltwater conditions or other chemical influences). Pipe lubrication fluid and delivery systems are designed to address these issues. As shown in Figure 2-5, volume-controlled pipe lubrication systems allow preprogrammed quantities of lubrication fluid to be dispensed based on anticipated ground conditions at predetermined locations along the DSPT alignment. Whereas uniform lubrication systems dispense a constant, predetermined volume along the length of the drive, regardless of ground conditions. Where monitoring of the annular pressure is required, a sensor is used to monitor and document the pressure of the lubrication fluid inside the annulus.

Figure 2-5. (left) Variable volume-controlled versus (right) uniform lubrication.
Source: Courtesy of Herrenknecht.

Figure 2-6. Operator controls.
Source: Courtesy of Laney Group.

The additional pipe lubrication that is introduced at the launch seal (Section 2.3.7) provides a resupply of lubrication fluid to the annulus. This resupply can be either through a static storage that allows a visual monitoring of the fill levels or through a dedicated low-pressure pump.

2.3.4 Control Cabin and Power Container

The control cabin (Figure 2-6) is usually located near the launch point so that the operator can observe the pipe thruster and other activities in the launch area. All the DSPT equipment components—such as the MTBM, pipe thruster, slurry pumps, lubrication system, and navigation system—are remotely controlled by the operator from inside the control cabin. All the individual functions of the equipment are operated through a central control panel.

2.3.5 Navigation and Guidance System

The MTBM follows a designed tunnel alignment and profile to install a pipe at the desired line and grade using a navigation and guidance system. Direct steerable pipe thrusting operates using a gyroscope, a hydrostatic water level, and a distance counter as primary navigation systems. These components are shown in Figure 2-7.

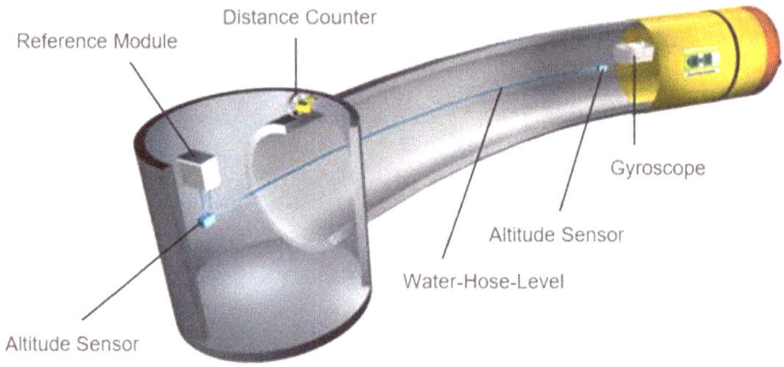

Figure 2-7. Schematic of DSPT primary navigation systems.
Source: Courtesy of Herrenknecht.

The gyroscope is placed inside the MTBM alongside other sensors. It provides information about the equipment's position and heading, yaw, pitch, and roll angle. These parameters are recorded and displayed in real time to the operator. The hydrostatic water level system provides data for vertical control, and the measuring wheel provides the installed length of pipe. These three systems working together provide the primary position control based on traditional navigation that begins with the point of origin, distance, and direction of travel over time.

The primary position control systems are calibrated through a control survey. Various possibilities exist to accomplish this survey and selecting the appropriate survey method depends on project-specific requirements and the diameter of the thrust pipe. The most precise approach is an optical survey through the pipeline to determine the position of the MTBM. Surface-based verification o f t he c urrent p osition i s a lso a n o ption, although the expected survey accuracy is less.

Most DSPT designs include a straight entry tangent section. This section can be used for an optical control survey because it offers a line of sight to the MTBM, allowing a surveyor to determine the position of a pre-established target on the back of the MTBM without entering the thrust pipe. This optical control survey is generally performed around 100 to 160 ft (30 to 50 m) after launching the MTBM either from the open end or through a window opening cut into the thrust pipe (Figure 2-8). The window opening approach is used if survey from the open end of the thrust pipe is not possible because of the length and orientation of the initial thrust pipe string. If the thrust pipe is intended to be the final product or carrier pipe, the section of pipe with the window opening is typically pushed out of the ground into the reception workspace and removed from the pipe string.

Figure 2-8. Window opening in a thrust pipe for survey access.
Source: Courtesy of GeoEngineers.

The DSPT design geometry, defined by the start and end points of straight and curved sections, together with the radii, and tangent/arc length segments, is entered into the guidance software prior to launch. The operator typically sees a visual representation of the alignment compared to the MTBM location, along with projections of where the machine is heading based on preceding information. The MTBM is steered by hydraulic cylinders at the articulating joint as it is advanced gradually in the desired direction.

2.3.6 Umbilical

The umbilical shown in Figure 2-9 is the aggregate of all service lines that carry drilling fluid feed and slurry return, lubrication fluid, water, compressed air, data communication, guidance, and electric and hydraulic power supply. The umbilical runs through the length of the thrust pipe, and it typically rests on suitable supports with rollers or casters to protect an interior lined pipeline and/or to facilitate the removal of the umbilical. For larger pipe diameters, the support frame may also include rails to transport personnel on a wheeled power sled.

To install the pipeline umbilical, a half-pipe of suitable length (Figures 2-10 and 2-11) is commonly cantilevered from the end of the thrust pipe by welding to or sleeving over the thrust pipe. The purpose of the half-pipe is to provide a platform for assembling the umbilical and consequently

Figure 2-9. Umbilical and transport rails inside a pipeline.
Source: Courtesy of Herrenknecht.

Figure 2-10 (left). Half-pipe with a slurry pump.
Source: Courtesy of GeoEngineers.

Figure 2-11 (right). Assembly of an umbilical on a half-pipe.
Source: Courtesy of GeoEngineers.

moving it step by step into the thrust pipe. The half-pipe is used to support a slurry pump during DSPT installation. Assembly of the umbilical should be such that the following criteria are met:

- Umbilical obstructs as little cross-sectional area as possible.
- None of the umbilical is in contact with the inner wall of the thrust pipe.

- All components of the umbilical are clearly identifiable and traceable.
- Umbilical has sufficient spacing from the pipe walls for the welding of subsequent thrust pipe sections.

The thrust pipe is prefabricated on site in one or more pipe string(s) with the umbilical preloaded in the pipe. There are two primary approaches of connecting the umbilical and welding the pipe string. The thrust pipe strings can be positioned with the umbilical separated some distance from the joint (in which case the ends of the umbilical are connected from the inside after the pipe is welded) or with the umbilical already connected (which requires additional consideration during the welding process to avoid heat damage to the umbilical). The method used depends on the length of the individual pipe strings and umbilical, as well as the thrust pipe diameter.

The umbilical extends out of the thrust pipe (Figure 2-12) and connects to the different power units, slurry separation plant, and the control cabin. The umbilical is routed back toward the launch pit either on the ground alongside the thrust pipe or on top of the pipe string (Figure 2-13) to keep an orderly arrangement on site. The lines making up the umbilical outside the pipe string are more flexible than those inside the pipe string to accommodate the pipe string advancing toward the launch shaft/pit and pipe thruster.

Figure 2-12. Umbilical exiting thrust pipe.
Source: Courtesy of Prime Drilling.

Figure 2-13. Umbilical on top of a pipe and a slurry circuit next to the pipe. Source: Courtesy of Herrenknecht.

The weight of the umbilical inside the pipeline depends on several factors such as the diameter of the feed, slurry and lubrication lines, number and size of slurry pumps, and number and size of power cables (size and weight of cables increase with tunnel length). As it relates to pipe handling, installation stresses, and umbilical removal processes, the umbilical weight should be evaluated based on a project-specific configuration.

When clearing out the thrust pipe after completing the installation (and disconnecting the MTBM from the thrust pipe), the umbilical can be pulled out on either end. Most often, the umbilical is pulled out from the launch side. Rollers or casters allow the umbilical to be moved in and out of the thrust pipe, while protecting the thrust pipe interior from damage. The DSPT profile needs to be considered when assessing required pull forces on the umbilical.

Depending on jobsite layout and equipment configuration, routing of the individual lines of the umbilical outside the pipe string varies. To keep the slurry circuit as short as possible, the separation plant is typically located half-way along the longest pipe string section, as shown in Figure 2-14. Typically, the lubrication fluid mixing and storage equipment is placed in the same area as the separation plant.

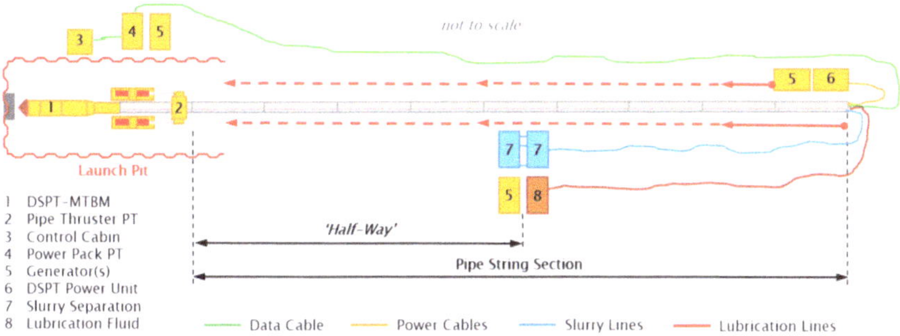

Figure 2-14. DSPT jobsite layout.
Source: Courtesy of Herrenknecht.

2.3.7 Launch Seal

A launch seal, as shown in Figure 2-15, serves as a barrier between the ground (at the launch point) and the launch shaft/pit. It holds back groundwater pressure, prevents outflow of lubrication fluid, and allows the lubrication fluid to be pressurized to stabilize the annulus. In conditions

Figure 2-15. Thrust pipe entering a launch seal (seal in the lower left-hand corner of the photo).
Source: Courtesy of Prime Drilling.

where groundwater is present, the lubrication fluid counterbalances hydrostatic pressure to prevent undesired flows into the annulus or launch shaft/pit.

Design of the launch seal should consider the lateral movement of the installed pipe inside of the seal to account for compression and relaxation during the push and regrip process. The design should include provisions for a possible partial or full retraction of the pipe string and MTBM. During retraction, the seals are prone to inverting back toward the launch shaft/pit, greatly reducing their effectiveness in holding back lubrication fluid and ambient groundwater pressure (Section 2.2.2). The launch seal must accommodate the outside diameter of the MTBM with its cutting wheel and the outside diameter of the thrust pipe.

2.3.8 Separation Equipment

DSPT utilizes an EDF/slurry loop to transport cuttings or excavation spoil to the separation plant at the ground surface. The separation plant screens the slurry to remove the excavated cuttings, which is then returned to the MTBM face as EDF. Depending on the makeup of the drilling fluid and the returned slurry, the processes involved in cleaning and recycling the EDF will vary.

The separation plant utilizes shakers and screens in combination with hydro-cyclones that separate the individual particles from the return slurry based on gradation. Silt and clay particles usually cannot be filtered out by traditional separation plants, and a hydro-cyclone may be required. Where silt and clay constituents are the predominant portion of the excavated ground, additional separation equipment such as centrifuges or filter presses can be added. For coarse sand and gravel, additional prescreens may be used. The entire separation process can be enhanced by introducing additives (polymers, surfactants, and clay inhibitors) to the EDF to keep materials aggregated for more efficient screening.

CHAPTER 3
PLANNING

3.1 INITIAL CRITERIA

Direct steerable pipe thrusting (DSPT) and other trenchless construction methods are suitable for conditions in which open-cut construction methods are not costeffective, are too disruptive, not permittable, or are not technically feasible. As outlined in Chapters 1 and 2, DSPT is typically used to install steel pipes of 30 to 60 in. (760 to 1,525 mm) in diameter, for distances up to and exceeding 5,000 ft (1,500 m), above or below the groundwater table in a range of geotechnical formations. Maximum achievable length is a function of subsurface conditions and the MTBM diameter. As the diameter of the MTBM increases, longer lengths and a greater variety of subsurface conditions can be accommodated. When trenchless crossings are required in unfavorable ground conditions or smaller diameters are necessary/preferred (bearing in mind that groundwater conditions influence method selection), other trenchless methods should be considered, such as

- Horizontal directional drilling (HDD),
- Pipe jacking,
- Microtunneling,
- Utility tunneling, and
- Pilot tube and other guided boring methods.

There are significant differences among these construction methods, both in application and in results. The engineer should review each of these methods to determine which can meet the project requirements and be used successfully under the specific ground conditions that will be encountered during the installation process.

Evaluation of the following parameters during the planning stages of a project will assist with determining the applicability and constructability of a potential DSPT project:

- Route/alignment;
- Installation lengths;
- Pipe diameter(s);
- Depth required;
- Permit requirements;
- Geotechnical conditions and geology, including groundwater;
- Potential for encountering buried objects (both natural and remnants of past construction) along the proposed alignment;
- Environmental conditions, including potential for ground or groundwater contamination;
- Line and grade requirements/tolerance;
- Product pipe material restrictions;
- Effects on surface activities;
- Location of existing, abandoned, or proposed underground utilities;
- Rights-of-ways (ROWs);
- Location of existing, abandoned, or proposed aboveground structures;
- Available workspace for equipment, pipe fabrication, handling, and staging;
- Site access; and
- Cost and construction schedule.

While considering suitable installation methods for a project, the engineer should consider all trenchless methods before concluding that the DSPT method is an appropriate construction method. Manuals and reports on engineering practice (known as *MOPs*) have been published by ASCE and others for different construction methods and are available for review.

If the engineer responsible for the project does not have a strong background in trenchless design, an engineering firm that specializes in trenchless design should be included in the design process early to make appropriate design recommendations.

3.2 ROUTING

Selecting the best route for a DSPT installation depends on a variety of factors. These considerations include avoiding known buried objects and working within existing or available ROW or temporary workspace, permit restrictions, access to workspace, geological conditions, and the existing topography.

3.2.1 Selection of Route

The route selection process is often dictated by the factors stated previously. One of the benefits of DSPT technology is the ability to provide routing that may not be available to alternate trenchless or open-trench construction methods. The route selected should reduce the need for relocating existing structures and/or utilities and provide adequate temporary workspace. The route should also take into consideration any required horizontal curves in the alignment. Introducing horizontal curves in a DSPT profile may lead to greater frequency of survey verification to achieve the required accuracy and possible increase in frictional loads, along with increased installation and operating stresses, particularly if the horizontal curves coincide with vertical curves.

3.2.2 Installation Distance and Diameter Range

For DSPT crossings, the achievable installation distance increases with pipeline diameter. The maximum distance is governed by the friction between the thrust pipe and the surrounding ground, the structural capacity of the thrust pipe, and the available thrust force.

The maximum pipe size currently available is 60 in. (1,525 mm). This is limited by current pipe thruster sizes and the thruster's clamping capability. Larger diameters are theoretically possible but have not yet been developed. With increasing diameter, the skin friction resistance and required push or thrust force increase.

The minimum diameter of DSPT is constrained by the internal components of the system. The smallest DSPT installed to date is 30 in. (760 mm); however, 36 in. (915 mm) is currently a more common lower limit. To allow for smaller pipe installations, equipment is being developed for pipe diameters of 20 in. (510 mm). The level of miniaturization and the performance capabilities of these components limit the maximum distance.

Figure 3-1 indicates approximate achievable installation distance for different ground conditions and pipeline diameters. Currently, for larger diameters, distances of up to 6,500 ft (2,000 m) are possible given favorable conditions.

The maximum distances for medium and small diameters are determined mainly by restrictions in the slurry circuit and the ability to generate enough power for excavation (i.e., rotating the cutting wheel). In addition, ground condition and type (and arrangement) of cutting tools is another limiting factor to consider when determining the achievable installation distance for a given pipe diameter. The smaller the diameter, the more difficult it is to deal with bedrock or boulders. In adverse ground conditions, penetration rates decrease and machine wear increases, both of which influence the maximum feasible installation distance.

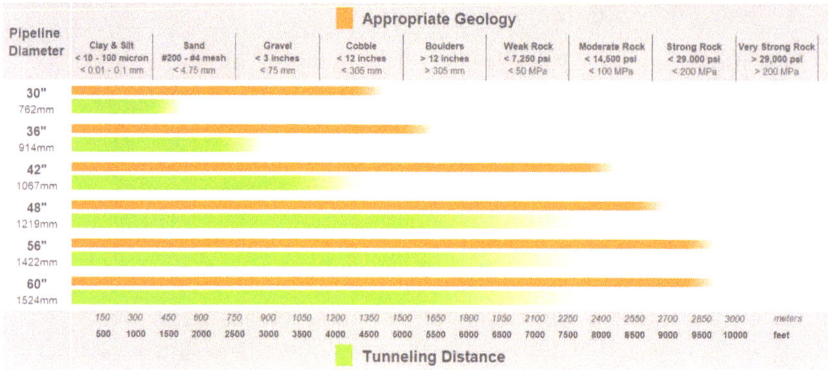

Figure 3-1. Installation distance and geotechnical considerations.
Source: Courtesy of Herrenknecht.

Because ground conditions vary around the world, it is not possible to provide fixed, reliable numbers on the subject of achievable installation distance or advance rates. Figure 3-1 is a general guide on what has been previously achieved, not a guarantee or limitation on what is possible.

3.3 PRELIMINARY GEOTECHNICAL FEASIBILITY

Prior to determining the feasibility of a DSPT project, a preliminary geotechnical investigation should be performed to evaluate the ground conditions. Other construction methods or routing alternatives should be considered if adverse ground conditions are identified early in the planning stages of a project. During the planning phase, published geology and historic explorations should be reviewed prior to conducting a detailed geotechnical assessment. The scope and timing of the subsurface investigation are discussed in further detail in Chapter 4.

3.3.1 Published Geologic Information

Published geologic information can come from a variety of resources. United States Geological Survey (USGS) and state geologic data can provide the nature and estimated depths of bedrock in some areas, and public water well information is also often available from various agencies. Other sources of historical data include aerial photographs, technical papers, historical documents, discussions with local contractors or agencies, and past project information or case studies.

3.3.2 Initial Site Visit

An initial site visit should be included in preliminary planning. During the site visit, the following should be evaluated:

- Visible surrounding geology, including rock outcroppings, natural erosion of riverbanks, and ground surface characteristics;
- Site access;
- Surface and groundwater constraints;
- Overhead and underground utilities;
- Site topography;
- Workspaces;
- Pipe laydown and stringing areas;
- Site preparation and clearing requirements;
- Adjacent properties;
- Environmental constraints;
- Urban constraints;
- Historic site usage;
- Haul routes;
- Disposal sites; and
- Freshwater availability.

3.4 PRELIMINARY GEOMETRIC DESIGN

The preliminary geometric design is influenced by the size of the feature being crossed, the radius of curvature limitations of the product pipe being installed, as well as the pipe thruster and MTBM capability, workspace constraints, topography, environmental constraints, and other features. A general starting point for the minimum radius of curvature (in feet) is to multiply the diameter of the pipe (in inches) times 100. The metric equivalent is the radius (in meters), which should be 1,000 times plus 10% greater than the diameter of the pipe (in meters). For example, a 36 in. (915 mm) diameter pipe should be designed with a minimum radius of 3,600 ft (1,100 m). This can vary based on the grade and thickness of steel being used. Such a method is usually satisfactory for most preliminary design layouts but should be refined during detailed design with pipe stress analyses, as discussed in Chapter 6.

An additional component of geometric design is a tangent section extending from the entry point, especially if the topographic or ground conditions require the use of a surface casing or conductor casing. Entry and exit angles are measured from the horizontal, and they are limited by equipment capability and site conditions. Entry angles should be designed to provide ease in breakover (the point of inflection in the pipe string

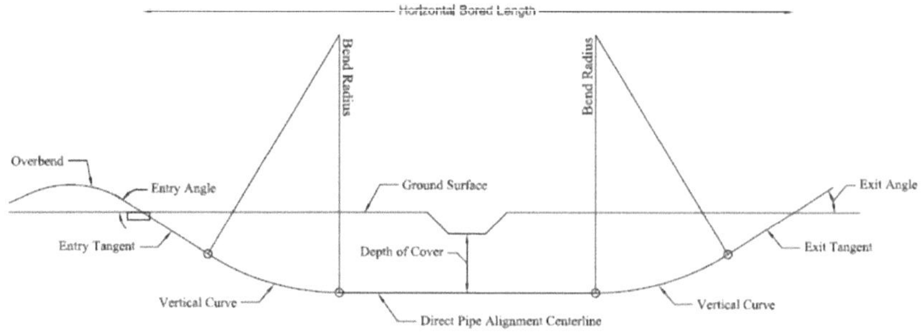

Figure 3-2. DSPT basic geometric components.
Source: Courtesy of Brierley Associates.

where the pipe transitions from upward to downward sloping) support for the pipe handling. Steeper entry angles require the pipe section breakover to be supported at an elevated position during the entire pipe installation process. Entry angles typically range from 4 to 10 degrees and should be modified to correspond with the equipment utilized (Section 3.4.2). Figure 3-2 shows the basic geometric components of a DSPT design.

3.4.1 Launch and Reception Shafts/Pits

Each side of the crossing should be carefully evaluated to determine the best launch and reception locations. If workspace or access is limited on one side of the crossing, often this ends up being the reception point. Topography also plays a large role in determining these points. Although a horizontal surface is ideal for equipment setup, the use of a gently ascending ground surface behind the launch area (as shown in Figure 3-4) will reduce or eliminate the need for heavy equipment to handle the pipe during installation. Pipe thrusting has its primary operations occurring from the entry side, with typically only MTBM retrieval and pipe tie-in occurring on the reception side.

If geometrical constraints require it, the MTBM is launched in a shored excavation or shaft. However, as discussed previously, it is often launched from a shallow pit (less than 1.5 m or 5 ft with no ground support) or from the ground surface. Either way, the launch area serves to anchor the pipe thruster/pusher. In addition, it is the staging area for all other operations and the hub of the entry side jobsite (Figure 3-3). See also Sections 3.1 and 3.5.1, as well as Chapter 8.

Launch shaft/pit designs range from a simple open pit without shoring (relying on ground strength or friction for stable side-sloped pit walls), to shallow shafts stabilized with sheet piling, to elaborate concrete and steel structures. The most common are open pits and sheet-piled shafts (Figures 3-4 and 3-5, respectively). The primary criteria influencing the launch shaft/pit design are geological and topographical conditions, presence and level of groundwater, size of staging area, and required or desired depth of launch (Section 3.4.4). The ground conditions are particularly important because they have a direct influence on the anchoring of the pipe thruster, which is the basis for all launch shaft/pit designs.

Figure 3-3. Example of a typical entry side jobsite, which includes surface casing, pipe thruster, assembled pipe string, control cab, crane, and generator.
Source: Courtesy of GeoEngineers.

Figure 3-4 (left). Open side-sloped launch pit with tented separation equipment in the background.
Source: Courtesy of Herrenknecht.

Figure 3-5 (right). Launch shaft with sheet piling ground support.
Source: Courtesy of GeoEngineers.

Each launch shaft/pit, even if the MTBM is launched from the surface, requires a foundation for the pipe thruster to transfer reaction loads into the ground (Figure 3-6). This foundation must resist the forces generated from pushing or pulling. Both direction and magnitude of the thruster forces are usually complex and require thorough consideration (Chapter 5). Depending on the geotechnical conditions and site constraints, the anchoring design becomes more elaborate in very soft or loose ground conditions.

The depth of the launch shaft/pit is governed by a placement of the launch seal (Chapter 2, Section 2.3.7), which is mounted to the entry wall and serves as the barrier to prevent groundwater, engineered drilling fluid, slurry, or lubrication fluid from spilling out of the annular space and entering into the launch shaft/pit. However, the design of the profile, desired entry angle, resulting overbend, and type of launch shaft/pit also influence the depth of the launch shaft/pit and the height of the entry wall. In situations where this height is not sufficient for a launch seal to be mounted, it is also feasible to raise the entry wall to an elevation above the surface level and then to add additional confinement above the pipe by placing artificial ground cover (temporary fill) in front of the entry wall and the launch shaft/pit (Figure 3-7; Section 3.4.3). Such shaft/pits, although economical, may

Figure 3-6. Anchoring structure for a pipe thruster consisting of pipe piles. Source: Courtesy of Herrenknecht.

Figure 3-7. Launch pit with an artificial cover.
Source: Courtesy of Herrenknecht.

require additional pipe handling during construction. The amount of cover at launch should be at least one to two times the outside diameter of the MTBM. This will depend on the type, quality, and permeability of temporary fill; disturbance; in-place density; risk and impact of inadvertent drilling fluid and slurry returns; and other project conditions.

In some cases, for instance when launching an MTBM from the surface, a conductor casing is buried or driven into the ground (Figure 3-8). This casing is somewhat larger in diameter than the MTBM, and it typically incorporates a welded launch seal. The casing may include a liner to reduce the risk of damage to the thrust pipe coating during installation. Thus, the conductor casing serves as an entry wall and artificial cover, while the MTBM rests inside the buried section of the casing.

In general, the reception consists of a small trench (or pit) excavated by a trackhoe or other construction machinery (Figure 3-9). Often, the reception

Figure 3-8. Conductor casing with a surface launch configuration.
Source: Courtesy of Michels Construction.

Figure 3-9. MTBM reception in an unsupported trench.
Source: Courtesy of GeoEngineers.

pit is sloped and follows the exit angle of the borehole, allowing the MTBM to punch out gradually to the surface. The pit should provide enough room for worker access to remove the MTBM from the pipeline and lift it away for transportation. The reception workspace must provide sufficient space for access for a mobile crane (or similar) and trucking and hauling equipment.

Where the reception occurs inside a structurally supported shaft, the exit wall for the MTBM is penetrated by the MTBM. This might be necessary on projects where required by ground surface topography, or when there is a risk of the lubrication fluid spilling from the bored hole's overcut in an unacceptable quantity into the environment, thus requiring a seal or other suitable measure to be installed. Any reception shaft must be planned to safely support the ground and groundwater conditions and provide access for workers and MTBM extraction.

3.4.2 Entry and Exit Angles

The entry and exit angles for a DSPT installation are comparable to angles used in other trenchless installation methods such as HDD. Entry and exit angles from 0 to 15 degrees are common, but steeper angles may be accommodated with site grading.

Steep entry angles allow the MTBM to reach depth in a shorter distance (see Section 3.4.3 for additional information); however, pipe handling and slurry removal during construction may require additional consideration. Steep entry angles generate larger (and higher in elevation) overbends that require more attention on the jobsite. Pipe handling and welding become more challenging with increasing overbend heights, and slurry removal may require increased pump capacity due to additional head pressure and must be dealt with accordingly (Figures 3-10 and 3-11). Conversely, shallow entry angles allow for a smooth transition of the pipeline from the stringing area into the bored hole, simplifying the on-site work process significantly (Figure 3-12).

Launch angles over 10 degrees are generally considered steep for DSPT applications, angles between 4 and 10 degrees are viewed as preferred, and angles less than 4 degrees are uncommon. The pipeline diameter, geotechnical conditions, ground surface topography, ability to pump the slurry to the surface, and the risk of inadvertent drilling fluid returns influence the angle. Typically, shallower entry angles are preferred with larger pipe diameters to reduce the overbend challenges associated with the size and weight of the thrust pipe.

For the exit point, the angle is typically of less concern. In most cases, the exit angle is rather shallow because the MTBM and pipeline are simply thrust gradually out of the bored hole onto the surface. This also provides for a more convenient removal of the MTBM from the thrust pipe.

Figure 3-10. Pipeline and overbend supported by pipe rollers on freight containers. Source: Courtesy of Michels.

Figure 3-11. Steep entry angle and pipeline supported by elevated side booms.
Source: Courtesy of Michels.

Figure 3-12. Shallow entry angle and smooth transition of a pipeline onto a surface.
Source: Courtesy of Michels.

3.4.3 Required Depth Based on Surface Features

The required depth of installation is dependent on a variety of factors. As a general practice, the cover should be a minimum of two times the diameter of the MTBM in use. This depth may need to be increased based on geotechnical considerations and the risk of surface heave, settlement, and surface release of fluid. Surface and subsurface features should also be evaluated. For instance, the risk of settlement underneath a gravel road typically carries less risk than that of a large four-lane highway. A systemic settlement analysis is typically performed after geotechnical information is available to assess settlement resulting primarily from collapse of the annular space overcut; it may also be necessary to evaluate settlement risks in the event the MTBM must be retracted during construction. Face and hydrostatic counterbalancing pressures are monitored during installation and controlled to allow for a balanced excavation.

Furthermore, current DSPT equipment available on the market has certain external pressure limitations that must be considered when planning a project. Currently, 4 bar (60 psi) sealing systems limit the overall vertical depth to approximately 130 ft (40 m) of hydrostatic pressure. For additional information, refer to Chapter 5, Section 5.2.5.

3.4.4 Required Cover and Overburden

Because DSPT incorporates one of the key features of microtunneling (relatively low drilling and lubrication fluid pressures, relative to HDD), the potential risk of formation fracturing or hydraulic fracturing (frac-outs) is correspondingly lessened. However, it is not eliminated. The pressure for drilling fluid and lubricant to fill the overcut is typically slightly higher than ambient pressure in the bored hole. This allows DSPT to be installed at shallower depths and through softer geological formations with relatively low hydraulic fracturing risk than that associated with HDD (Figure 3-13). As a rule of thumb for DSPT, the recommended depth of cover is between two and five times the pipeline diameter. Depending on the ground conditions and other design considerations, the depth of cover may be greater.

The excavated overcut is typically smaller in DSPT relative to HDD but larger than in microtunneling. The excavated bored hole diameter is often less than 10% larger than the pipeline diameter, whereas in HDD, the borehole diameter is, in general, 30% to 50% larger than the pipeline diameter. In microtunneling, the radial overcut is typically less than 2 in. (50 mm). Over time (after completing the installation), the overcut or annular space around the installation pipe, typically filled with drilling fluid and bentonite lubricant, will continue to fill with overburden ground that collapses into the annular space (systemic settlement). Accordingly,

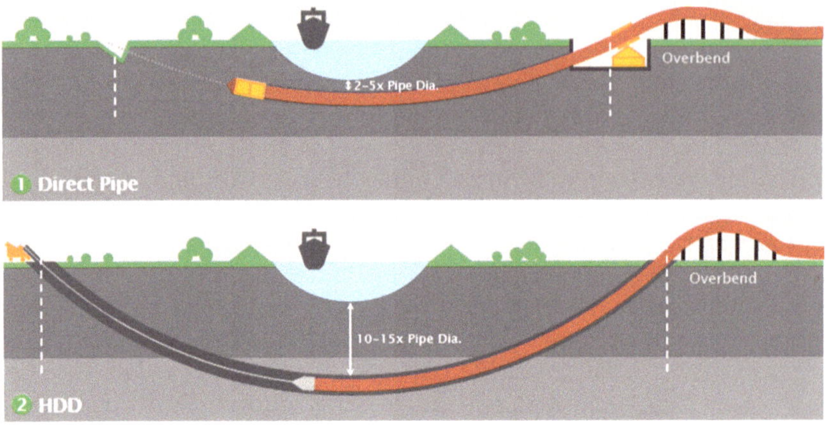

Figure 3-13. Comparison of cover, depth, and distance for DSPT and HDD based on drilling fluid pressures.
Source: Courtesy of Herrenknecht.

smaller annular space overcuts generally result in lower magnitudes of ground surface settlement (Chapter 5, Section 5.3.3).

3.4.5 Uphill–Downhill Considerations

Most DSPT projects have launch and reception points near the same elevation. However, shore approaches, for instance, tend to have an increased elevation difference because the launch point is on land and the reception point is on the seabed. In many cases, although, this elevation difference is moderate (unless the shore approach is on a coastal cliff). However, some projects are located on hillsides or across river valleys with significant elevation differences such as those shown in Figures 3-14 and 3-15. In general, elevation differences may become critical when approaching or exceeding the pressure rating of the DSPT equipment, which varies by the MTBM. The following is intended to elaborate on some of the considerations for projects with significant elevation differential.

In microtunneling, there is a general tendency to push uphill on drives with large elevation differences. The general logic is, should there be any kind of leakage through the system and into the pipe, it will run downhill out of the tunnel instead of accumulating behind the MTBM. In DSPT, however, there is often a vertical curve with a low point in the alignment somewhere along the bore path. Hence, fluids can only accumulate behind the MTBM in the declining portion of the tunnel profile.

On an uphill drive in microtunneling (installing a straight bore), one consideration is always the pipe change when the jacking frame is retracted

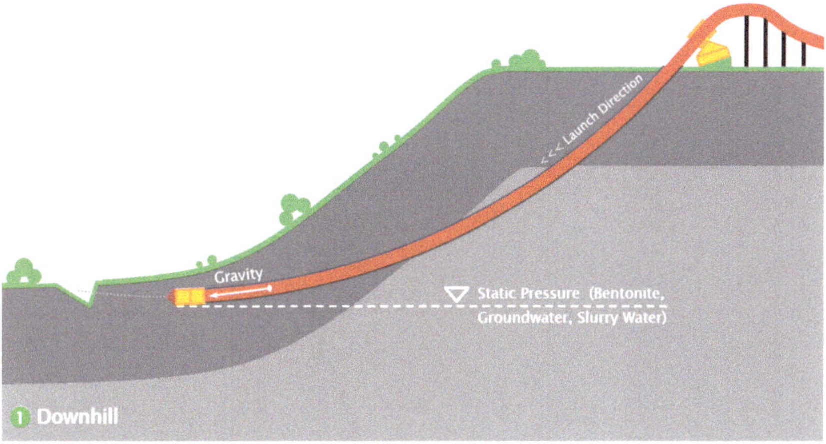

Figure 3-14. Downhill drive.
Source: Courtesy of Herrenknecht.

to give way for the next pipe. At this moment, there is a potential for the pipe to move backward into the launch shaft, which makes using a pipe brake (pipe clamp) mandatory. However, a DSPT installation from the low to the high side still typically requires an initial declining profile before transitioning to an inclined profile from the low point to the surface (Figure 3-15). Downward force from the pipeline moving backward occurs only after the alignment climbs up past the point of equilibrium. To move backward, the pipe has to overcome friction along the arched profile and any friction above the point of equilibrium. Therefore, an uphill drive is typically a good choice, but if the pipe can move backward, a second pipe thruster/pusher can be added to serve as a pipe brake.

Another aspect to consider is the launch seal and the lubrication in the overcut. A downhill drive does not require a robust launch seal, because the lubrication fluid rests inside the tunnel under nominal pressure and does not burden the seal with high lubrication pressure in the annular space. For an uphill drive, in theory, the launch seal would be subjected to the static hydraulic pressure of the lubrication fluid caused by the elevation difference. Practical experience, however, shows that this is not always the case. Friction between the lubrication fluid and the pipeline and tunnel wall dissipates some of the static pressure.

For slurry circulation, an uphill drive is typically favorable because the added static pressure differential that develops from gravity after crossing the point of equilibrium (or approximate elevation of entry) assists the slurry return circuit when approaching the elevation of the reception point. However, the slurry returned from an uphill drive enters the separation equipment at the full pressure head of the elevation difference. Depending

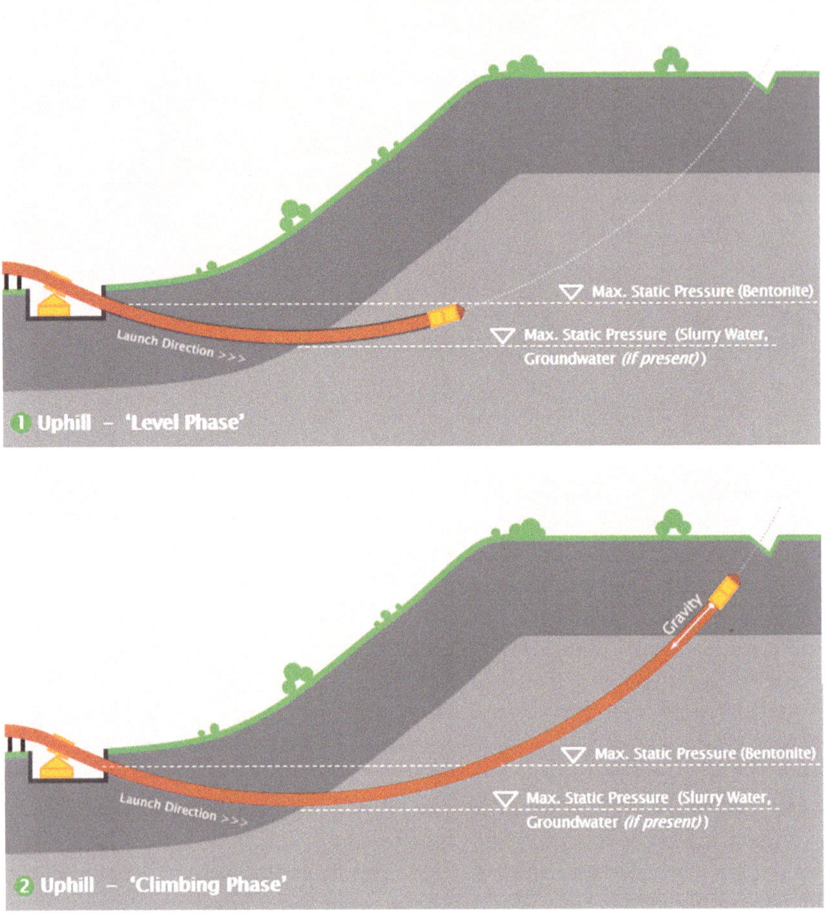

Figure 3-15. Uphill drive.
Source: Courtesy of Herrenknecht.

on the elevation difference, if the pressure is too high, it could overwhelm the mechanical systems and piping in the slurry circuit. Conversely, on a downhill drive, the slurry must always be pumped uphill, requiring additional pumping capacity.

Another consideration is the internal seals of the MTBM and of the drilling fluid and slurry components (Chapter 5, Section 5.2.5). In a downhill drive, the MTBM and fluid components face the full pressure head between the launch point elevation and the lowest point in the tunnel alignment. In general, there are two different pressures to be considered: (1) the static pressure on the outside of the MTBM and on the outside of the pipeline caused by the column of lubrication fluid in the annular space or pipe annulus, and (2) the internal pressure on the slurry circuit (not accounting for groundwater). Regardless of the pressure levels, the equipment (the MTBM and slurry components) must be sealed to prevent internal systemic leakage within the MTBM.

3.4.6 Shore Approaches

Shore approaches have become a specialty application for DSPT. Shore approaches require additional considerations related to water depth and associated pressures, environmental considerations, and recovery of the DSPT equipment. For the most part, the entire operation is land-based. However, when exiting, a number of issues arise that need to be addressed beforehand and carefully planned. In general, when the MTBM is detached from the thrust pipe, the following must be considered:

- Is it a dry or wet recovery?
- How are pressures inside the pipe or casing equalized?
- Is flooding of the thrust pipe acceptable?
- Is release of fluids into the waterbody acceptable?
- How will the umbilical lines be removed?
- How will the MTBM be protected and removed?
- What are the weather window and offshore equipment schedules?

For a dry recovery, many of the aforementioned considerations are not applicable. In a dry recovery, the entire pipe string with the MTBM attached in front is either lifted out of the water onto a vessel, platform, or offshore lay barge, or the MTBM and pipe enter a dry shaft or cofferdam. Then, the MTBM is separated from the thrust pipe. For this recovery method, it is important that the necessary hoisting capacity is available offshore. Lay barges normally have ample lifting capacity, but the timing must carefully planned.

As far as the DSPT equipment is concerned, there are two things to consider when planning a dry recovery. In general, the equipment removed consists of the MTBM and several following articulated sections. Lifting all these sections while they are connected would require individual support of the sections or a lifting beam to stabilize the entire MTBM unit (Figure 3-16). Typically, the equipment is not designed to be cantilevered in front of the thrust pipe. Therefore, lifting points and lifting method must be chosen carefully. Furthermore, the umbilical lines, when disconnected from the MTBM, will need to be pulled back toward the launch pit and recovered. A decision has to be made on whether to retrieve the umbilical lines before or after commencing dry recovery. On a dry recovery, the umbilical lines could theoretically be pulled out of the thrust pipe toward the recovery point, but that may not be efficient over water.

Figure 3-16. Wet recovery of the MTBM.
Source: Courtesy of Herrenknecht.

A wet recovery is different. In a wet recovery, divers disconnect the MTBM from the thrust pipe underwater. Tides, waves, surf, visibility, and weather play a significant role in a wet recovery. Also, water depth is an important factor for bringing in a jack-up barge or pontoon to provide hoisting capacity. An alternative to hoisting at the exit location is to raise or float the MTBM to the surface with air-inflated bags commonly used in diving and salvage operations and then to tug the MTBM to the nearest harbor or nearby shore location to take advantage of land-based cranes and hoisting equipment.

The most important aspect of a wet recovery, however, is equalizing pressures. Pressure outside the MTBM and the thrust pipe is defined by the water depth, whereas pressure inside the pipe is atmospheric pressure. To equalize these pressures, a portion (or all) of the thrust pipe must be flooded to the same level as the water body *before* disconnecting the MTBM. Depending on the thrust pipe (e.g., in the case of a product pipe with interior coating), the client may not allow the pipe to be flooded with water. If the thrust pipe cannot be flooded, a cap or bulkhead must be preinstalled behind the MTBM to seal it off toward the landside. This bulkhead allows for the pressure equalization required to permit removal of the MTBM without flooding the thrust pipe. In all wet recoveries, the MTBM's internal components must be protected from flooding when equalizing pressures with the use of a bulkhead installed at the tail of the MTBM. After the pressures are equalized, divers disconnect the MTBM and trailing cans from the thrust pipe and the MTBM and trailing cans are hoisted to the surface.

3.5 WORKSITE LAYOUT

Worksites are configured using different sizes and shapes based on the length of the crossing, available length for pipe sections, contractor preference, and size and type of thrusting equipment. Every worksite is unique, but one typical preference for the layout is having the workspace offset from the centerline of the installation. This allows optimized staging of equipment and enables personnel to work efficiently when the pipe string becomes suspended and/or impedes ROW access. The worksite layout should be designed to accommodate a smooth traffic flow through the site for approximately 20 to 30 tractor-trailer loads of DSPT equipment; additional thrust pipe loads; and ancillary equipment, personnel, and supply access. For additional information, refer to Section 3.4.1.

3.5.1 Standard Workspace Size

A typical workspace configuration for DSPT should consider three primary components of the operation: the pipe thruster, the pipe string, and the auxiliary equipment. The pipe thruster is situated close to the launch point and can be located either on the surface or in a shallow shaft/pit, depending on the profile, topographic, and worksite conditions. The pipe thruster is usually paired with an operator control cabin, as well as a hydraulic power pack. This typically requires an area approximately 50 ft wide ×100 ft long (15 m × 30 m) or larger if possible. The length and width of the pipe stringing and fabrication area will vary depending on the length of the crossing and available space. In general, this workspace needs to accommodate the pipe string fabrication and pipe handling equipment. The design workspace should be wider and can be shorter if more than one pipe string is required. In addition to the pipe thruster and pipe string, auxiliary equipment will need to be placed near the pipe string, in addition to the slurry handling and separation facilities (which are ideally located near the midpoint of the pipe string layout). This equipment setup is flexible, depending on the contractor, but usually requires additional temporary workspace.

3.5.2 Modifications Based on Site Limitations

Similar to HDD or microtunneling, the operational setup can often be modified to accommodate jobsite conditions. One of the primary restrictions often encountered is the amount of space required for one continuous pipe string. The DSPT method allows for the utilization of short pipe strings or the entire length of the crossing depending on space constraints. Auxiliary equipment can also be located to accommodate jobsite restrictions such as site access or environmentally sensitive areas.

3.5.3 Pipe Fabrication and Storage Area Requirements

Direct steerable pipe thrusting allows for multiple pipe fabrication scenarios. The industry has utilized pipe strings as long as approximately 2,500 ft (760 m) or as short as 40 ft (12 m), although typically, a pipe string of more than 120 ft (35 m) is desirable for economic reasons. In general, a pipe stringing and fabrication area workspace width of at least 100 ft (30 m) is desirable, although the width may need to be increased depending on the number of pipe strings. It is preferable to fabricate the entire pipe string prior to the start of DSPT operations if sufficient space is available.

3.5.4 Pipeline Handling and Support

Handling and supporting the thrust pipe and umbilicals can be achieved in many ways and depends on many factors such as site conditions, easements, site access, as well as available equipment. One of the traditional methods for handling overbend and the thrust pipe itself is by cranes and sideboom cranes. As the pipeline travels steadily into the excavated hole, sideboom cranes require some additional attention because they must follow the forward motion of the pipeline. This process is relatively slow compared to open-trench or HDD applications because pipe is thrusted into the bored hole as it is being excavated. When setting up for another section of the pipeline, if required, having sideboom cranes on the jobsite can be beneficial, instead of other lifting or hoisting equipment. However, for handling the overbend, sideboom cranes are not always practical as they do not easily adjust as is sometimes required. Furthermore, the reach and height that they can attain is limited and, depending on the entry angle, other means of pipeline support must be considered for the over bend.

For large and/or highly elevated over bends, either mobile telescopic cranes or traditional lattice crawler cranes are used. The latter is the more widespread and found on many construction sites today.

An alternative that reduces the manned support equipment required is a stationary support structure such as a fill slope or empty freight containers on which pipeline rollers are placed to support the overbend during pullback (Figure 3-10). Beyond the overbend section, pipeline rollers are used most often to support the length of a pipeline or section as it is prefabricated and laid out on the jobsite for installation using DSPT.

3.5.5 Site Access

Continuous site access for large trucks is necessary during DSPT operations to deliver additional supplies and fresh water and to dispose of spoils. Where possible, access should be provided utilizing the shortest distance from improved roads to reduce costs. Quite often, remote sites or surface conditions require the construction and restoration of temporary access roads or the placement of surface mats for the equipment and pipe laydown areas.

3.6 PRELIMINARY INSTALLATION COST CONSIDERATIONS

A variety of factors need to be taken into consideration to develop a cost estimate for a pipe installation using DSPT. Preliminary cost estimation is challenging partly because of the relative newness of the technology, the number of contractors who are capable of installing pipelines by DSPT, and the number of completed crossings. Direct and indirect costs should be considered when evaluating a potential project. Some of the indirect costs to consider are risk mitigation, environmental effects, and social costs to the surrounding community.

3.6.1 Cost Considerations

Cost is an important consideration when comparing trenchless construction methods. Typically, DSPT costs are compared to either microtunneling or HDD. However, DSPT has benefits not found in either of these other methods that should be considered when evaluating construction options. Although direct costs should not be the only consideration, it is often a good starting point for cost comparisons and for deciding the most appropriate trenchless option. Some other benefits that should be considered include reduced life cycle costs, traffic effects, scheduling requirements, and reduced environmental impact.

The direct cost of a DSPT installation is influenced or controlled by several factors. These factors include the following:

- Geotechnical conditions and geology;
- Installation pipe diameter and wall thickness;
- Buried objects along the alignment;
- Groundwater;
- Number of competing projects and number of qualified bidders;
- Labor conditions;
- Complexity of the project (including alignment and grade requirements);
- Location;
- Depth, size, and complexity of shafts, if required;
- Length of installation;
- Mobilization, setup, and moving costs;
- Direct steerable pipe thrusting penetration rates;
- Spoils handling and disposal;

- Special requirements, schedule, and/or restrictions;
- Project owner's reputation; and
- Availability of DSPT equipment;

3.6.1.1 Geotechnical Conditions and Geology. For crossings in clayey ground, costs may increase as the clay content increases because it takes longer to excavate and process spoils. Costs are usually higher for excavation in rock because of slower advance rates and an increase in equipment wear. Larger-diameter machines fare better with rock than do smaller machines; however, larger-diameter machines generate more spoils and require larger-diameter pipes. Because larger machines excavate more ground per foot of advance, they may require larger slurry lines and separation plants to support faster penetration rates. Larger equipment typically results in higher costs.

3.6.1.2 Installation Pipe Diameter and Wall Thickness. Currently, there is a range of pipe diameters that are appropriate for DSPT installations. Designs that stray from the range of diameters that have corresponding pipe thrusters will increase the bid price, and thicker walled pipe will result in higher bids because of increased cost to produce. In general, as the diameter of the installation pipe increases, the cost of the installation increases as well.

3.6.1.3 Buried Objects along the Alignment. Hitting one or more buried objects can slow or stop the forward progress of the MTBM and cause delays. Delays mean increased construction time, which, in turn, increases costs. When buried objects are encountered, additional costs are incurred for the construction of rescue shafts, re-tunneling an alternate alignment, or for the use of alternative tunneling methods. Even buried objects that do not stop the MTBM can damage equipment or accelerate wear.

3.6.1.4 Groundwater. In general, DSPT is a preferred construction method when high groundwater head is encountered, because dewatering along the alignment is not needed. Costs of pumping and disposal of nuisance water increase with increasing shaft depth.

3.6.1.5 Number of Competing Projects and Number of Qualified Bidders. The number of projects under construction, recently awarded, and in the bidding phase affects contractor backlog. Contractors who have no backlog or a minimal backlog may set their bid prices lower. As contractors' backlog increases, they can be more selective about which projects to bid on and the prices set for the work. The level of a project's risk, whether real or perceived, is critical to contractor interest in bidding. The number of qualified bidders also affects costs. Increased competition tends to reduce bid prices.

3.6.1.6 Labor Conditions. The location of the project, and whether the area is subject to union rates or prevailing wage rates, can influence the number of workers and labor costs.

3.6.1.7 Complexity of the Project. As the complexity of the project increases, so do the project's risks and pricing. In addition to geotechnical conditions and other factors, the geometric installation requirements may drive costs up, particularly if very tight tolerances are required. Increased complexity may also reduce the number of bidders and increase costs.

3.6.1.8 Location. If a project is located close to any established DSPT contractor, the mobilization costs may be lower. If a DSPT contractor is on a local project, then mobilization costs are lower for other projects in the area if similar equipment can be used. Likewise, if a project is in a remote location, prices tend to escalate.

3.6.1.9 Depth and Size of Shafts. The cost of a shaft incrementally increases with its depth and size, presence of groundwater, issues related to rock versus soil conditions, and complexity of site access. Shaft configuration may also increase cost.

3.6.1.10 Length of Installation. Longer drives are normally more expensive because of the equipment required, including slurry pumps, ventilation, and guidance systems. Other nonequipment cost considerations include production efficiency, the strength of the jacking pipe, and survey requirements. For crossings with drive lengths and in ground conditions illustrated in Figure 3-1, the unit cost per foot typically decreases as the total footage of an installation increases.

3.6.1.11 Mobilization, Setup, and Moving Costs. A significant cost of a DSPT project is the equipment mobilization. Mobilization costs include preparation of the cutter wheel with new cutting tools, shop testing, refurbishing of ancillary equipment, and transport of the equipment to the jobsite. Depending on the diameter of the cutter wheel, power generation needs, and the type of slurry separation equipment required, approximately 20 to 30 tractor-trailer loads will be needed. A setup of the DSPT equipment and the spoils separation plant can take up to 2 weeks or more, depending on the complexity of the project.

3.6.1.12 Direct Steerable Pipe Thrusting Penetration Rates. The cutter wheel's ability to progress through the ground affects the project costs. In general, a soft-ground cutter wheel penetrates faster through sand than clay. Penetration rates for a rock cutter wheel also decrease in clay.

Depending on the cutter wheel's configuration and size, encountering gravel, cobbles, and boulders can lower penetration rates.

3.6.1.13 Spoils Handling and Disposal. Slurry requires processing. Costs increase with increased size and complexity of the slurry separation plant. The factors that affect plant size include the following:

- Properties of the slurry to transport cuttings,
- Efficiency of the separation plant in removing the fines content,
- Amount of the slurry that can be reused,
- Requirements for additional on-site storage, and
- Time required to separate the spoils.

All spoils generated at the site require disposal. Increasing the distance to the final disposal site increases cost. Contaminated ground or groundwater also increases the cost of disposal.

3.6.1.14 Special Requirements, Schedule, and/or Restrictions. Any special requirements imposed on a project increase the project's costs. Owners and engineers should work with third parties to minimize such special requirements. Typical special requirements or restrictions may include the following:

- Limits on workspaces;
- Overall project schedule requirements;
- Limits on work hours;
- Clearance of road surfaces during peak commute hours;
- Limits on truck trips, pipe delivery, or spoil removal;
- Limits on noise; and
- Dust control and frequency of road cleaning.

3.6.1.15 Project Owner's Reputation. An owner may have a reputation regarding contracting requirements and risk sharing. This reputation can increase or decrease a contractor's contingencies on a project and their willingness to bid.

3.6.1.16 Availability of Suitable Direct Steerable Pipe Thrusting Equipment. The diameter and capabilities of the DSPT equipment and the availability of machinery for purchase or rent may be a cost consideration.

3.6.2 Indirect Social Effect Costs

The indirect social effects of a construction method should be factored into the selection process of the method. The cost of indirect social effects has been documented in Boyce and Bried (1994, 1998), Vickridge et al. (1992), and Sterling (1994). Properly conveying the duration and impacts before construction can reduce the costs of mitigating indirect social impacts during construction or prevent social impacts from becoming direct costs.

Indirect or social effects that may be considered include the following:

- Vehicle traffic disruption,
- Pedestrian disruption,
- Disruption to business activities,
- Loss of tax revenue,
- Disruption to residences,
- Carbon footprint,
- Noise,
- Dust,
- Odors,
- Restoration costs,
- Public safety, and
- Effects caused by land disturbance required (i.e., environmental costs).

For a detailed discussion of these costs, refer to the ASCE Microtunneling Standard 36 (2015).

3.6.3 Environmental Costs

The effect of a construction method on wetlands and environmentally sensitive areas should be considered in the evaluation and selection of construction methods. Direct steerable pipe thrusting can significantly reduce the construction effect on wetlands and other environmentally sensitive areas.

The effect of construction through contaminated ground and groundwater should be taken into account where potentially contaminated ground and/or groundwater have been identified during the investigation process. A contingency plan identifying methods for handling contamination may be required as part of the contract documents when construction is to be completed through potentially contaminated areas. Where possible, excavations should be located outside the contaminated area or constructed using methods designed to reduce the risk of migration of contaminated ground and groundwater. Direct steerable pipe thrusting below the level of the contaminated material, especially through impervious layers, can significantly reduce or eliminate the problems associated with DSPT operations through these areas. Launch and receiving pits should be located as far as possible from critical environmental features, stream banks, and wetland buffers. It is important to perform a hydrogeological assessment to inform the owner, engineer, and contractor about potential groundwater issues that could affect water flow to wells, springs, seeps, or other groundwater expressions.

3.6.4 Risk Analysis

Depending on the owner's risk tolerance and assessment practices, a formal risk identification and mitigation process may be used to document and quantify project risk. A risk register is a common tool for this purpose and may be started for the project during the planning phase. The risk register is a living document and should be updated throughout the project's life. The engineer, with input from the owner, should identify and analyze the risks associated with viable construction methods. It may be helpful in some cases to conduct an alternative analysis of DSPT and other trenchless methods that accounts for both initial and long-term direct costs, indirect costs, and the effect of construction on third parties.

The process for creating a risk register includes the following activities:

- Holding a brainstorming session with key members of the design team, owner, and contractor (if applicable) to identify project hazard and risk scenarios for each construction method proposed;
- Grouping the hazard and risk scenarios into related categories (e.g., excavations, DSPT installation, public outreach, permitting, property acquisition, safety, and contractual issues);
- Identifying the causes of each identified hazard or risk scenario;
- Evaluating the probability or likelihood that a specific hazard or risk scenario will occur;
- Evaluating the potential effect of the occurrence in terms of relative cost and/or scheduling delay to the project;
- Rating the hazard and risk scenarios on their likelihood and severity;
- Taking mitigation measures to reduce the probability and/or severity of hazard or risk scenarios;
- Re-rating the hazard or risk scenarios with mitigation measures incorporated;
- Updating the risk register with newly identified hazard or risk scenarios as the project advances in design and into construction.

As an example, the severity of the effect and the probability of a risk are both ranked from 1 to 5. Each risk ends up with a ranking between 1 and 25 (Table 3-1). Risks ranked in the red zone should be mitigated, if possible, during design into the yellow or green zone. Risks ranked in the yellow zone should be mitigated during design into the green zone, to the extent possible. The risk register becomes a project management tool for making sure that mitigation measures, once identified, have been incorporated into the design.

Table 3-1. Risk Table with Severity of Impact and Probability of Occurrence.

Probability impact	Rare (1)	Unlikely (2)	Possible (3)	Likely (4)	Most likely (5)
Very high (5)	5	10	15	20	25
High (4)	4	8	12	16	20
Moderate (3)	3	6	9	12	15
Low (2)	2	4	6	8	10
Very low (1)	1	2	3	4	5

3.6.5 Contingency Costs

Unknown or unforeseen conditions are associated with any underground project. Contingency costs should be identified and included in any construction cost budgets prepared by the owner. Typical risks are evaluated based on cost, schedule, safety, and environment. The risks can be included in different categories such as ground conditions, permitting, contracting, and shaft construction.

Despite possessing a complete risk register, some risks cannot be eliminated or reduced by design or construction mitigations alone. To allocate these risks, the owner should consider adding contingencies to the construction cost budget. These contingency costs can be held outside the construction contract or included as allowances within the bid schedule. Allowances are fixed dollar amounts, set by the owner, to cover additional items for the project that could not be quantified at the time of the bid. As an example, allowances can be included for disposal of contaminated ground.

When owners show their willingness to identify and share project risks, bidding contractors are more likely to include fewer contingencies in their bids. See Chapter 7 for more discussion on dispute resolution as a method available to allocate and share risks on a project.

REFERENCES

ASCE. 2015. *Standard design and construction guidelines for microtunneling*. ASCE 36-15. Reston, VA: ASCE.

Boyce, G. M., and E. M. Bried. 1994. "Benefit–cost analysis of Direct Pipe in an urban area." In *Proc. No-Dig 94, Ann. Conf. of the North American Society of Trenchless Technology*, Dallas, Texas.

Boyce, G. M., and E. M. Bried. 1998. "Social cost accounting for trenchless projects." In *Proc., No-Dig 98, Ann. Conf. of the North American Society of Trenchless Technology*, Albuquerque, New Mexico. 3–12.

Sterling, R. L. 1994. *Indirect costs of utility placement and repair beneath streets.* Rep. No. MN/RC-94/20. St. Paul, MN: Minnesota Dept. of Transportation.

Vickridge, I., D. J. Ling, and G. F. Read, 1992. "Evaluating the social costs and setting the charges for road space occupation." In *Proc., No-Dig 92, Ann. Conf. of the North American Society of Trenchless Technology*, Paper B2.

CHAPTER 4

DETAILED SITE INVESTIGATIONS AND GEOTECHNICAL ENGINEERING

4.1 INTRODUCTION AND CHAPTER SUMMARY

This chapter addresses site investigations that are important for planning, constructability/feasibility evaluation, detailed design, construction, long-term operability, and resiliency of direct steerable pipe thrusting (DSPT) projects. These site investigations include the following:

- Geotechnical assessment and characterization,
- Utility and other structure surveys (both aerial and subsurface),
- Topographic and bathymetric survey,
- Environmental conditions, and
- Seismic conditions.

It is not the intent of this chapter to provide detailed guidance on conducting all site investigation programs or laboratory and field testing for specific DSPT projects. Each project is unique and should be evaluated individually. A site-specific investigation should be performed by a qualified DSPT design engineer and geo-professional.

4.2 SITE INVESTIGATION

It is common practice to conduct a detailed site investigation in phases. Each subsequent phase of the study should address gaps identified in the prior phase of the investigation and provide the information needed to plan and execute the next phase of study.

After the development of a conceptual DSPT design for the site, the first phase of site investigation typically consists of a desktop study to identify the anticipated site conditions along the preliminary alignment including accessibility and preferred boring locations where gaps in geotechnical data exist. The next phase consists of ground verification or aerial reconnaissance to verify the desktop study and geotechnical studies to follow. The next phase typically consists of conducting geotechnical explorations, further characterizing the subsurface materials, evaluating the properties of the encountered geotechnical materials, and performing various laboratory tests on selected samples (as discussed subsequently). Depending on project needs, a portion of the test borings may be converted to observation wells to characterize seasonal or long-term groundwater table changes.

Using the results of the site investigations, a working model can be developed for the crossing location. This working model includes the anticipated subsurface geotechnical materials and their properties, stratigraphy, and groundwater conditions. Data from observation wells can allow the interpretation of groundwater information to establish seasonal or tidal fluctuations of groundwater levels. Groundwater samples obtained from the wells can also be analyzed for water quality and corrosivity, assuming that the wells are installed in accordance with current environmental procedures.

For some projects, further refinement of the working model may be required where significant gaps or uncertainties are noted. A subsequent phase of investigation represents an opportunity to test the validity of the subsurface model. Depending on the outcome of this work and the importance of specific conclusions relative to subsurface conditions, it could be prudent to conduct an additional phase of even more detailed investigations to resolve remaining ground condition uncertainties. In addition to client-specific guidelines, numerous references discuss geotechnical investigations (Mayne et al. 2002, PRCI 2015, Parker 2004).

4.2.1 Desktop Study

A desktop study typically consists of the review of existing or publicly available information, including

- Geologic maps;
- Local or regional groundwater (and surface water) conditions and historic highs and lows;
- Current and historical aerial photography;
- Historical maps and photographs;
- Existing or historical subsurface information (which may include boring logs of subsurface investigations, water well records, environmental investigations, or other well or historical completion data);

- Construction records from prior nearby projects or projects in similar subsurface conditions; and
- Records of potential contamination, including a review of state and private tracking tools and record searches as well as aerial photography, mapping, and photography.

Depending on the scope of the project, the results of the desktop study may be published as a preliminary geotechnical report or memorandum or included in a preliminary design report. For some projects, the results of the desktop study may simply be used to plan the primary-level subsurface investigation and may be included with the primary geotechnical reporting.

4.2.2 Geotechnical Subsurface Investigation Methods

Initial site investigation work, which is typically conducted during the planning or preliminary design stage of a project, should include a site visit and a review of material gathered during the desktop study.

The most common form of subsurface exploration is the sampled geotechnical boring. This involves advancing a small-diameter vertical borehole using a casing or hollow-stem auger and collecting samples of the material encountered using a soil or rock coring sampler at specific depth intervals. An important first decision is to determine how many borings to drill, their locations, and termination depths. Although it is common to complete land-based geotechnical programs for DSPT installations, water-based programs are less common but should be considered based on the distance between shorelines and/or where high-risk geotechnical conditions are known to exist. Many guidelines exist to aid in the development of the geotechnical program and include suggestions on a number of borings. At a minimum, borings are typically necessary near proposed launch and reception locations. Depending on the variability of the geology and the availability of historical data, additional borings along the alignment at a spacing of approximately 500 ft (150 m) to 1,000 ft (300 m) may be appropriate (Mayne et al. 2002). This range is a general guideline and subject to site-specific conditions, client and regulatory input, the crossing length, and the geotechnical engineer's judgment.

Geotechnical borings should extend sufficiently below the conceptual vertical alignment to identify changes in the strata that could affect the construction and lead to a deeper than anticipated installation. Borings at the entry location will be needed to design thrust anchorage configuration, and excavations, shoring, and dewatering (if deemed necessary for the project). Typically, explorations should extend at least 20 ft (6 m) below the proposed conceptual alignment. If soft, compressible soils are present at or near the proposed pipe profile, some of the borings should be deepened.

The final decision on the number, type, and depth of borings should be based on the importance and complexity of the project, anticipated crossing risks, installation length, site access, and the nature and complexity of the ground conditions. More extensive programs should be developed for higher risk or more complex installations.

During completion of the test borings, soil samples are commonly obtained using split spoon samplers as part of a standard penetration test (SPT). The reader should refer to the most recent version of ASTM D1586 (ASTM 2019) for further information. For samplers and hammers that deviate from ASTM D1586 standards, such as those of ASTM D3550 (ASTM 2018), the reader is cautioned that these methods require correction for a variety of conditions and should not be directly used for engineering design.

Continuous samples are recommended through the proposed DSPT profile and for one pipe diameter above and below the profile. Outside of these zones, samples typically should be taken at intervals no greater than 5 ft (1.5 m) and at changes in geologic strata. Additional samples may be required for thruster foundation and shoring engineering of launch and reception workspaces.

If bedrock is encountered, continuous coring should be completed using triple-tube (e.g., HQ, NQ) core barrels. The length of the core run will depend on the rock mass quality but, in most cases, will be on the order of 5 ft (1.5 m). Coring may also be a viable sampling method if oversized materials (e.g., cobbles, boulders) are encountered.

Additional field tests may include in situ vane shear testing, hand penetrometer testing, point load testing on rock samples, pressure meter testing, and in situ permeability tests. Refer to the applicable ASTM Standards for details.

For some crossings, other forms of geotechnical investigation may be prudent to supplement the geotechnical boring program. Test pits, rotosonic borings, and large-diameter drill holes can be useful methods of subsurface investigation, especially for identifying the presence of gravels, cobbles, and boulders; where subsurface conditions are extremely variable, or in urban settings in which areas of undocumented fill may exist. They may also provide direct evidence of the type and distribution of objects that exist in the ground and reveal important facts about possible ground behavior characteristics such as sloughing, squeezing, and groundwater inflows. Cone penetration tests (CPTs) may be useful in displaceable ground. Cone penetration tests involve pushing a piezocone probe into the ground and can be a cost-effective method of obtaining a continuous record of subsurface conditions. Cone penetration tests should always be calibrated at the site with one or more coupled soil borings, and their use is contingent on soil density/hardness and CPT rig capacity.

There are many geophysical techniques that may be effective in interpolating subsurface conditions between the geotechnical borings. The applicability of geophysical techniques should be evaluated on a case-by-case basis. A geophysical professional should be consulted during planning of the geophysical program to ensure that the appropriate methods are employed.

Another key element of the subsurface investigation program is continuous on-site monitoring of the work by a trained geo-professional communicating with the engineer. Thorough observation and documentation of field activities and drill rig behavior must be provided so that the results of the field work can be accurately and correctly interpreted.

4.2.3 Utility and Other Structure Surveys

It is important that the locations of existing utilities be reliably established as early as possible during planning to evaluate the feasibility of implementing the DSPT method. All utility information (aerial and subsurface) should be collected in connection with the project alignment(s). ASCE 38, *Standard Guidelines for the Collection and Depiction of Existing Subsurface Utility Data* (ASCE 2002) provide a reference for utility surveys. Quality Level B and Quality Level A utility data should be obtained especially if the utility expectancy is complex and/or risky. Note that using a state's on-call service can only result in Quality Level C or Quality Level D information.

During the planning and design phase, it may be possible to make changes to the alignment and entry/exit locations to avoid conflict and the necessity for a potentially costly and time-consuming utility relocation. If conflicts cannot be avoided, relocation plans should be prepared. Relocation of the utilities can be done in advance of the project or as part of the project and can be of a temporary or permanent nature. The utility owner must be engaged as early as possible for input and review. Utilities should be shown on the contract drawings. Information such as utility quality level, diameter, depth, material type, owner, trenched cross section, and backfill material used (according to ASCE 38, ASCE 2002) should be identified if available.

Utilities that will remain in place and are near the project may require protection and should be monitored during construction. Evaluation of settlement and heave risks may be appropriate depending on the type and age of the utility, clearances, and ground conditions.

4.2.4 Topographic and Bathymetric Survey

Project survey data are arguably the most important data on the project, especially beneath bodies of water, levees, and in urban areas where vertical alignment may need tight control. It is important to understand the source of survey data, the accuracy of the data, the horizontal and vertical datum(s) used, and the requirements of the client and/or regulatory agencies early during project planning.

Horizontal and vertical control work should, at a minimum, comply with accuracy standards as defined in the Federal Geographic Data Committee (FGDC) Geospatial Positioning Accuracy Standards, Part 4: Standards for Architecture, Engineering, Construction (A/E/C), and Facility Management (FGDC 2002). The project team must define the horizontal and vertical accuracy of data to be collected. In addition, the subsequent map accuracy standard should also be defined by *class*, in accordance with the American Society of Photogrammetry and Remote Sensing Accuracy Standards (ASPRS 2014). The mapping class must be compatible with the data accuracy needs.

The US Army Corps of Engineers (USACE) also has standards (in the form of engineering manuals) that address topics such as photogrammetric and light detection and ranging mapping, monuments and markers, and the use of global positioning systems (USACE 2007, 2011, 2012, 2015). Survey work should be completed under the direction of a licensed land surveyor or in accordance with local surveying practice, and all aerial mapping work should be completed under the supervision of an ASPRS-certified photogrammetrist.

Similarly, accuracy for bathymetric surveying must be defined prior to data collection. Global Positioning System (GPS) data must be tied to the sonar or other remote sensing data collection devices and calibrated for changing river levels/tides; GPS should employ real-time kinematic methods to establish relative positional accuracies relative to a published coordinate system and horizontal and vertical datum based on the higher-accuracy National Geodetic Survey control network to more effectively tie the overwater surveying into the project control points.

Project survey datums must be defined and be compatible with the needs of the project such as permit requirements and with client needs and preferences. Conversion between coordinate systems and datums as a matter of procedure should be avoided if possible. If not avoidable, the conversion calculations to a known/published coordinate system and datum (both horizontal and vertical) should be documented.

4.3 GEOTECHNICAL ASSESSMENT AND CHARACTERIZATION

It is important that ground and groundwater conditions and their range of variability be reliably established as early as possible during project planning to evaluate the feasibility or constructability of using DSPT methods to complete a specific crossing. If the ground conditions and anticipated behaviors are deemed feasible for the DSPT approach, the information obtained from the site investigations will be used to design and construct the project. However, if high-risk geotechnical conditions and/or difficult site constraints exist, it may not be feasible to use DSPT as a construction method.

4.3.1 Characterization and Subsurface Risk

DSPT is considered a closed-face trenchless method and is applicable for use in a variety of ground conditions, including those that classify as flowing or running (according to the terminology used in Table 4-1). Squeezing and swelling ground can result in elevated thrust forces, which may exceed the capacity of the chosen equipment. Additives can be included in the lubrication fluids to reduce squeezing tendencies. Identification and mitigation of the risks associated with these ground types can be key to a successful completion of a drive.

As previously mentioned, the presence of high-strength and high-abrasivity clasts may result in challenges to a successful installation. For a discussion of additional risks associated with slurry microtunneling that are applicable to DSPT, see ASCE Standard 36 (ASCE 2015).

Section 4.2.2 describes site characteristics that are important to identify and quantify during the site investigation, as they may affect the overall feasibility of the DSPT installation. Refer to Chapter 5 for additional discussion of feasibility assessment. Ground conditions are typically explored through a comprehensive exploration program and laboratory testing program.

For cohesionless soil (e.g., sand and gravel), the engineering properties that are typically identified include the following:

- Grain size distribution,
- Unit weight,
- Hydraulic conductivity,
- Density (typically in terms of SPT *N*-values),
- Maximum clast size and percentage of material (i.e., gravel, cobbles, and boulders if present),
- Abrasivity of soil and strength and abrasivity of rock clasts, and
- Estimation or measurement of the internal friction angle.

For cohesive soils (e.g., clay), the engineering properties that are typically identified include the following:

- Moisture content,
- Atterberg limits (plasticity index and liquid limit),
- Swelling potential,
- Unit weight,
- Shear strength,

Table 4-1. Tunnelman's Ground Classification.

Classification		Behavior	Typical Soil Types
Firm		Heading can be advanced without initial support, and final lining can be constructed before ground starts to move.	Loess above water table; hard clay, marl, cemented sand and gravel when not highly overstressed.
Raveling	Slow raveling ----- Fast raveling	Chunks or flakes of material begin to drop out of the arch or walls sometime after the ground has been exposed, due to loosening or to over-stress and "brittle" fracture (ground separates or breaks along distinct surfaces, opposed to squeezing ground). In fast raveling ground, the process starts within a few minutes, otherwise the ground is slow raveling.	Residual soils or sand with small amounts of binder may be fast raveling below the water tale, slow raveling above. Stiff fissured clays may be slow or fast raveling depending upon degree of overstress.
Squeezing		Ground squeezes or extrudes plastically into tunnel, without visible fracturing or loss of continuity, and without perceptible increase in water content. Ductile, plastic yield and flow due to overstress.	Ground with low frictional strength. Rate of squeeze depends on degree of overstress. Occurs at shallow to medium depth in clay of very soft to medium consistency. Stiff to hard clay under high cover may move in combination of raveling at excavation surface and squeezing at depth behind surface.
Running	Cohesive - running ----- Running	Granular materials without cohesion are unstable at a slope greater than their angle of repose (approx 30° -35°). When exposed at steeper slopes they run like granulated sugar or dune sand until the slope flattens to the angle of repose.	Clean, dry granular materials. Apparent cohesion in moist sand, or weak cementation in any granular soil, may allow the material to stand for a brief period of raveling before it breaks down and runs. Such behavior is cohesive-running.
Flowing		A mixture of soil and water flows into the tunnel like a viscous fluid. The material can enter the tunnel from the invert as well as from the face, crown, and walls, and can flow for great distances, completely filling the tunnel in some cases.	Below the water table in silt, sand, or gravel without enough clay content to give significant cohesion and plasticity. May also occur in highly sensitive clay when such material is disturbed.
Swelling		Ground absorbs water, increases in volume, and expands slowly into the tunnel.	Highly preconsolidated clay with plasticity index in excess of about 30, generally containing significant percentages of montmorillonite.

* Modified by Heuer (1974) from Terzaghi (1950)

- Compressibility/overconsolidation ratio,
- Consistency [typically in terms of unconfined compressive strength (UCS) or SPT *N*-values],
- Grain size distribution sieve and hydrometer analysis, and
- Maximum clast size and percentage of material (i.e., gravel, cobbles, and boulders if present).

The laboratory testing program should be assigned by a geo-professional with knowledge and experience in assessing the properties of the various ground conditions encountered in the subsurface explorations with input from the DSPT design engineer. Laboratory testing should be done in accordance with ASTM standards for each test.

4.3.2 Soil Behavior

The Tunnelman's Ground Classification System, first described by Terzaghi (1950) and refined by Heuer (1974), describes ground behaviors and their effect on larger, conventionally constructed tunnels through soil. The system is also a useful tool for the evaluation of soil behavior as it relates to DSPT. Descriptions of various ground behaviors, in terms of the Tunnelman's ground classification system, are presented in Table 4-1, which can be evaluated for its effects on DSPT. General ground behaviors include raveling, running, flowing, squeezing, and firm ground. Details and descriptions of various ground behaviors and their effects on small-diameter tunneling, which are also directly applicable to DSPT, are presented in the ASCE Microtunneling Standard 36 (ASCE 2015).

4.3.3 Bedrock Conditions

When the subsurface conditions identified during the desktop study phase of a project indicate that the DSPT alignment will be located partially or entirely in bedrock, the exploration program should address key physical properties of the bedrock. Design considerations associated with bedrock crossings are discussed in detail in Chapter 5.

The bedrock properties that should be evaluated during the field portion of the subsurface exploration program include the following:

- Depth and extent of bedrock,
- Bedrock type (e.g., lithology),
- Core recovery,
- Rock quality designation (RQD),
- Weathering,
- Discontinuities (frequency, spacing, orientation, aperture, infill material, roughness, and shape),
- Presence of fracture-controlled groundwater, and
- Field hardness.

Conventional rock coring potentially accompanied by downhole acoustic/optical televiewer logging is another way to obtain many of the field data outlined previously and also allows the user to collect in situ data when discontinuities and orientation could influence design.

After the completion of the subsurface exploration program, a laboratory testing program should be completed on select samples obtained in the field, to properly evaluate the following properties. Some useful tests include

- Hardness;
- Unconfined compressive strength (UCS);
- Modulus of elasticity;
- Mineralogy and Mohs hardness (e.g., thin sections);
- Abrasivity (e.g., Cerchar); and
- Slake durability, jar slake, and swelling tendencies for claystone and shale.

When collecting rock cores, they should be documented with their depth indicated, and photographs with a scale and color guide immediately upon retrieving the sample. Once the sample is retrieved or moved, it will begin to change from its in situ condition because of exposure to the atmosphere, vibration, and moisture loss.

Weathering index, fracture frequency, joint characteristics, and overall rock mass quality are also important for estimating boreability. Estimation of block sizes can be an important factor for evaluating the potential for blocks to become wedged in the cutter head openings and cutting chamber or to become separated or dislodged above the pipe and cause gouging of the pipe.

Rock quality designation is widely used as an indicator of overall rock quality. The designer should consider the ability of the bedrock in front of the face to be ripped, chipped, and broken into smaller pieces. This parameter is controlled by the frequency of discontinuities of the rock. This is typically evaluated based on RQD and the following three tests performed on intact samples of rock: UCS, hardness, and abrasivity. The properties of the intact rock are determined by conducting a series of laboratory tests. These factors also affect the tooling efficiency and survivability. A high percentage of siliceous mineral content (e.g., quartz) typically results in increased abrasivity of the rock, which may result in accelerated rates of tooling wear.

4.3.4 Groundwater

The groundwater level in unconfined aquifers and piezometric levels in confined aquifers can be estimated by installing observation wells and/or piezometers adjacent to pit locations. The designer may consider adding piezometers at intermediate points for longer installations. In addition, the hydraulic conductivity of water-bearing strata should be evaluated.

Groundwater conditions will have a significant influence on ground behavior (Table 4-1) and may influence machine operation and entry configuration. Hydraulic conductivity of water-bearing strata can be estimated using grain size correlations or borehole permeability tests.

4.3.5 Special Conditions

4.3.5.1 Gravels, Cobbles, and Boulders. The extent, frequency, size distribution, maximum clast size, and physical properties such as strength and abrasivity of gravels, cobbles, and boulders should be evaluated. If rock fragments are encountered within the sampler during the geotechnical investigations, steps should be taken to evaluate if the rock fragments are naturally occurring gravel or were generated by the sampler encountering and fracturing cobbles and/or boulders. Where there is a potential for gravels, cobbles, and boulders, large-diameter boreholes or rotosonic borings can be useful in obtaining representative samples for grain size analysis and determining the size and distribution of the gravels, cobbles, and boulders. Samples of the cobbles and boulders as well as the matrix material should also be obtained and tested.

4.3.5.2 Potential Buried Objects. Buried objects may impede or prohibit the advance of the MTBM. The likelihood of buried objects, their nature, and relative sizes should be established by the desktop study and site investigation. To avoid multiple attempts and possible settlement resulting from the retraction of the MTBM, the potential for buried objects and an understanding of their nature should be evaluated early in the planning phase to reduce the risk of encountering an obstruction. Naturally occurring buried objects may include buried trees, boulders, and floating corestones of unweathered rock. Man-made buried objects may include existing utilities, wells, or foundations.

4.3.5.3 Mixed-Face Conditions. Mixed-face conditions occur when there are distinct variations in ground conditions within the cross-sectional area of the face, such as soil over rock. Because potential tunneling difficulties can arise from mixed-face conditions, they should be avoided or reduced to the extent possible during the planning and design phases.

Mixed-face conditions can reduce steering control and present significant challenges to alignment, grade control, and stability of the face. Mixed-face conditions can also introduce the risk of overexcavation and stoppage of the machine's advance.

If mixed-faced conditions cannot be avoided, better line-and-grade control can be achieved in some cases by advancing the MTBM from hard ground into softer ground, reducing the advance rate, and steering into the hard ground of the mixed face to maintain line and grade. Allowance and adjustments to the construction schedule may be required.

For additional guidance, refer to the Microtunneling Standard 36 (ASCE 2015).

4.3.5.4 Mixed-Reach Conditions. Mixed-reach conditions occur when there are distinct variations in ground conditions along the alignment, such as soil and rock. Because potential tunneling difficulties can arise from mixed-reach conditions, they should be avoided or reduced to the extent possible during the planning and design phases.

During the design phase of the project, consideration should be given to matching the cutter wheel and tooling for the anticipated ground conditions. If the majority of the drive will be in rock, this could result in using a hard rock cutter wheel through the near-surface soil before encountering the rock. This could result in slow progress and difficult steering until the rock is encountered. If the rock is relatively soft, a mixed-face cutter head, equipped with both rock discs and soil scrapers, could be used.

Mixed-reach conditions often accompany mixed-face conditions and similar risks apply. For additional guidance, refer to the Microtunneling Standard 36 (ASCE 2015).

4.3.6 Applicability of Method Based on Subsurface Conditions and Classification

Because DSPT is considered to be a closed-face or full-face trenchless method, it is applicable for use in a variety of ground conditions, including those that classify as flowing or running (according to the terminology used in Table 4-1). Squeezing and swelling ground can result in elevated thrust forces, which may exceed the capacity of the chosen equipment. However, unlike microtunneling installations where the annular space is typically confined to control settlement, DSPT installations have been completed with greater annular spaces (although not as great as those typically associated with HDD installations) to overcome risks associated with squeezing ground. In addition, additives can be included in the lubrication fluids to minimize squeezing tendencies. Identification and mitigation of the risks associated with these ground types can be key to a successful completion of a drive.

As previously mentioned, the presence of high-strength and high-abrasivity clasts may result in challenges to a successful installation. For a discussion of additional risks associated with slurry microtunneling, which are applicable to DSPT, see ASCE Standard 36 (ASCE 2015).

4.3.7 Geotechnical Reports

All subsurface and reference data collected during the geotechnical desktop study and site investigations, professional interpretations thereof, and design and construction considerations should be summarized in project reports. A geotechnical data report (GDR) typically contains all the factual geotechnical information for the project, including explorations, laboratory and field-testing results, and geophysics and historical geotechnical data. The GDR typically does not include interpretations or recommendations. The GDR should be included in the contract documents.

Geotechnical design memoranda and/or a geotechnical interpretive report (GIR) should be prepared to present summaries of the geotechnical data, interpretation of the data, earth pressures to be used for design, discussions of the expected behavior of the ground, and other geotechnical design recommendations. Because the GIR and design memoranda are typically prepared prior to design, there is information contained in these memoranda and reports with respect to the project that may not be applicable at the time of bid. These reports may or may not be included in the contract documents at the owner's discretion.

Another standalone geotechnical report that may be prepared for a given project is a geotechnical baseline report (GBR). The GBR is more commonly used for municipal installations and is typically prepared as the design is being completed with the owner's input. The GBR serves as the definitive geotechnical baseline for use in the resolution of disagreements, disputes, or claims relating to differing subsurface conditions. The GBR presents contractual interpretations of the data to be used for bidding and construction, as well as baselines of expected ground behavior and other geotechnical construction considerations [see ASCE's Geotechnical Baseline Reports for Construction: Suggested Guidelines (ASCE 2007)]. If a GBR is produced for a project, it should be included in the contract documents. More information on GBRs, contract documents, and dispute resolution is provided in Chapter 7.

4.4 ENVIRONMENTAL CONDITIONS

Environmental conditions can be naturally occurring or the result of human activity. Naturally occurring conditions include hydrocarbons and asbestos as well as active or dormant landslides, active erosional features, and corrosive soils. Some human-created conditions include places of historical and cultural significance as well as landfills and contaminated ground and groundwater. Environmentally sensitive areas can include the following:

- Sources of water supply (wells, aquifers);
- Islands and island corridors;
- Beaches;
- Dunes;
- Wetlands and wetland transition areas;
- Breeding, nesting, and spawning areas;
- Migratory stopover areas;

- Wintering areas;
- Prime fishing areas;
- Migratory pathways;
- Water areas supporting submerged vegetation;
- Shellfish harvesting waters;
- Forest areas;
- Habitat for federal and state endangered or threatened plant and animal species;
- Federal and state wilderness areas;
- Areas designated as wild, scenic, recreational, or developed recreational rivers; and
- Cultural resources.

Properly planned projects can reduce or eliminate the effect of construction on these environmental conditions by crossing under the features.

Existing site features that could be affected by construction and DSPT operations should be identified during the site investigation. Historic buildings and environmentally sensitive areas usually require evaluation on a case-by-case basis to ensure protection. Effects to the surface or the near-surface environment can be reduced by locating entry and exit locations outside the sensitive area and drilling under it. In this case, the risk of impacts is reduced by simply passing under the sensitive area.

For contaminated ground (either human-made or naturally occuring), if the affected area cannot be avoided during planning-level route selection, the volume of contaminated spoils generated can be reduced by locating the entry and exit locations outside the contaminated area and then advancing the bore through the affected area. In this case, the volume of contaminated ground is reduced to the excavated bore volume.

An environmental professional should be consulted for assistance in the development of contract requirements related to such issues as handling and disposing of the spoils, decontamination of equipment before and after construction, and personnel health and safety. It should be noted that the availability of contractors qualified to perform work at contaminated sites may be reduced.

4.4.1 Contaminated or Hazardous Ground or Groundwater

Encountering ground and/or groundwater contamination has health and safety, cost, and schedule effects on projects. Hazardous conditions can include hydrocarbons, asbestos, and other chemicals occurring naturally or as a waste by-product from human activities. Careful planning and execution may be effective to mitigate these impacts. Evaluation of the potential for encountering hazardous substances should be completed during the planning phase. If the likelihood of contaminants is high, then evaluation of the nature and extent of contaminants, if present, should be undertaken during the site investigation.

4.4.2 Flood Zones

Floods, storm surges, and flash flood conditions should be considered when identifying and selecting entry/exit locations for DSPT alignments during design. Designing for potential high-water conditions can reduce the risk of flooding of shafts/pits, reduce damage to DSPT equipment, prevent pollution from petroleum products, mitigate cost and schedule effects, and protect personnel. In addition, locating work areas within designated flood zones may require additional permitting effort, which can delay design activities.

Assessment of flooding risk during planning should include a review of information from Federal Emergency Management Agency, the National Oceanic and Atmospheric Administration (NOAA), and USACE, as well as local knowledge to determine if storm impacts are probable, along with expected seasonality and frequency of storm occurrence. In some locations, flooding may last for weeks and affect the project schedule. The ground conditions may become saturated and can affect the shafts/pits and DSPT operation.

If flooding conditions are a significant risk, the design may need to consider the following:

- Planning construction operations to avoid periods with a high potential for flooding;
- Designing shafts/pits to address the full hydrostatic pressure condition or plan to temporarily flood the shaft/pit to eliminate unbalanced hydrostatic pressures;
- Designing for fully saturated ground conditions;
- Temporarily raising the site grade, if permittable, to provide better access;
- Designing for increased uplift that may occur during flooding or determining maximum groundwater level and method of monitoring that is allowed before flooding occurs;
- Developing plans for evacuation and noting that access roads may flood before the project site floods;
- Protecting or removing sensitive equipment, oils, and fuels;
- Developing plans to protect the project site from flooding when not occupied, including weekends and holidays;
- Developing plans to monitor water levels in advance and during large rainstorm events;
- Designing shafts/pits, pipes, and operations that prevent destabilization of flood control levees and embankments and recognizing that localized levee slope failures can occur during floods;

- Developing start-up plans after a flood event. Such plans should include mobilization of pumps; disposal of water from the pit; and cleanup of the surface, pit walls, and equipment within the pit.

In areas prone to flooding, the owner should consider including allowances or contingencies in the contract to compensate the contractor for this event, associated cleanup, and loss of time to the construction schedule. A major flood event may be considered a differing site condition or a force majeure event. It is recommended that flood events be defined in reference to specific water surface elevation and recurrence intervals (i.e., 1 in 100 years) and to identify the anticipated conditions in the contract documents.

4.5 SEISMIC CONSIDERATIONS

Seismic conditions including ground shaking and fault offset and related ground responses should be considered when selecting entry/exit locations, pipeline alignments, and design requirements for the entry/exit pits. Equally important are the anticipated postseismic event level of service and the consequences of failure for the pipeline (i.e., flooding or gas release and fire). The seismic considerations include protection of personnel and minimization of damage from fault offset, ground shaking, landslide, liquefaction, lateral spreading, and tsunamis.

Assessment of seismic risk should include a review of information from the United States Geological Survey, applicable state geological surveys and departments of transportation, and local knowledge to evaluate the location, nature, extent, recurrence interval, and design magnitude of seismic conditions relative to the project. In areas prone to seismic events, the engineer should consider the need for ground improvement and the fact that a project could be affected and delayed if an earthquake occurs during construction.

For additional information on seismic design of pipelines, consult O'Rourke (2003) and the Technical Council on Lifeline Earthquake Engineering Publications and Monographs, ASCE.

REFERENCES

ASCE. 2002. *Standard guidelines for the collection and depiction of existing subsurface utility data*, ASCE 38. New York: ASCE.

ASCE. 2007. *Geotechnical baseline reports for underground construction: Suggested guidelines*. Technical Committee for the Underground Technology Research Council. New York: ASCE.

ASCE. 2015. *Standard construction guidelines for microtunneling*, ASCE 36. New York: ASCE.

ASPRS (American Society of Photogrammetry and Remote Sensing). 2014. *Positional accuracy standards for digital geospatial data, version 1.0*. Bethesda, MD: ASPRS.

ASTM International. 2018. *Standard practice for thick wall, ring-lined, split barrel, drive sampling of soils*, ASTM D3550. West Conshohocken, PA: ASTM.

ASTM. 2019. *Standard test method for standard penetration test (SPT) and split-barrel sampling of soils*, ASTM D1586. West Conshohocken, PA: ASTM.

FGDC (Federal Geographic Data Committee). 2002. *Geospatial positioning accuracy standards part 4: Standards for architecture, engineering, construction (A/E/C) and facility management*, FGDC-STD-007.4-2002. Reston, VA: FGDC.

Heuer, R. 1974. "Important ground parameters in soft ground tunneling." In *Proc., Specialty Conf. on Subsurface Exploration for Underground Excavation and Heavy Construction*, 41–55. New York: ASCE.

Hung, C. J., J. Monsees, N. Munfah, and J. Wisniewski. 2009. "Technical manual for design and construction of road tunnels—Civil elements." FHWA-NHI-09-010. Washington, DC: FHWA.

Mayne, P W., B. R. Christopher, and J. DeLong. 2002. "Subsurface investigations—Geotechnical site characterization." Rep. No. FHWA-NHI-01-031. Washington, DC: FHWA.

O'Rourke, M. J. 2003. "Buried pipelines." Chap. 23 in *Earthquake engineering handbook*, edited by C. Scawthorn and W.-F. Chen. New York: CRC Press.

Parker, H. W. 2004. "Planning and site investigation." In *Proc., Tunnelling. 1st Congresso Brasileiro de Túneis e Estruturas Subterrâneas Seminário Internacional South American Tunnelling*. [In Portugese.] Sao Paolo.

PRCI (Pipeline Research Council International). 2015. "Installation of pipelines by horizontal directional drilling, an engineering design guide." Rep. No. CNST-1-3. Houston: PRCI.

Terzaghi, K. 1950. "Geologic aspects of soft ground tunneling." Chap. 11 in *Applied sedimentation*, edited by P. D. Trask. New York: Wiley.

USACE (US Army Corps of Engineers). 2007. "Control and topographic surveying." *Engineering manual*. EM 1110-1-1005. Washington, DC: USACE.

USACE. 2011. "Navstar global positioning system surveying." *Engineering manual*. EM 1110-1-1003. Washington, DC: USACE.

USACE. 2012. "Survey markers and monumentations." *Engineering manual*. EM 1110-1-1002. Washington, DC: USACE.

USACE. 2015. "Photogrammetric and LiDAR mapping." *Engineering manual*. EM 1110-1-1000. Washington, DC: USACE.

CHAPTER 5
ENGINEERING DESIGN

5.1 INTRODUCTION

The design process for a direct steerable pipe thrusting (DSPT) installation draws from both horizontal directional drilling (HDD) and microtunneling. This chapter discusses the following topics as they relate to the engineering design of DSPT installations:

- Geometric design,
- Thrust force estimate,
- Hydraulic fracture evaluation,
- Overcut and settlement,
- Slurry and lubricating fluid considerations,
- Launch and reception area considerations, and
- Acceptable geometric tolerances.

Design of the thrust pipe is covered in Chapter 6.

5.2 GEOMETRIC DESIGN

The geometric design for a DSPT installation typically consists of five basic elements:

- Launch tangent (at the design entry angle),
- Launch curve,
- Bottom tangent,
- Reception curve, and
- Reception tangent (at the design reception angle).

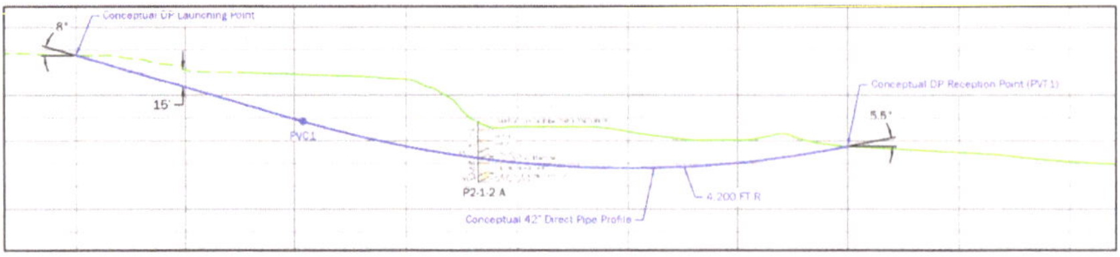

Figure 5-1. Conceptual DSPT shore approach.
Source: Courtesy of GeoEngineers.

Depending on project constraints, additional elements could be added such as horizontal or compound curves. Similarly, it may be possible to eliminate one or more elements. For example, the bottom tangent could be removed, as it is not necessary for a successful installation because of the ability to continuously monitor the MTBM's location and induce steering inputs simultaneously to maintain the design alignment during DSPT operations. An example DSPT shore approach concept profile is shown in Figure 5-1. In this figure, the entry tangent extends from the launching point to the point of vertical curvature 1 (PVC1); note the absence of bottom and exit tangents sections in this relatively short design.

5.2.1 Launch and Reception Angles

Launch angles for DSPT installations often range from 4 to 8 degrees below horizontal but could be as steep as 15 degrees or as flat as 0 degrees. Selection of a launch angle should consider the required installation depth along an alignment over a horizontal distance, available setback distance from a crossing feature to achieve the required depth of cover, topography, product pipe diameter, and equipment necessary to transition the pipe into the thruster clamps and launch window. As launch angles increase, a greater emphasis and consideration for the breakover section is required. Reception angles often vary depending on the scope of the installation and geometric constraints. Reception angles for shore approaches range from 2 to 6 degrees from horizontal where land applications range from 4 to 10 degrees but are only limited by the logistics of removing the MTBM from the reception point. As with the launch angle, the reception angle can be modified as needed to suit project constraints.

5.2.2 Radius of Curvature

The design radius used to establish the geometry of the installation should be selected based on product pipe diameter, ability to steer the MTBM within the anticipated ground conditions, and the potential for abrasion to the thrust pipe. Similar to HDD design for steel pipe, typical design bend radii (in feet) are on the order of 100 times the outer diameter of the thrust pipe (in inches), although a higher or lower ratio can be used based on the results of pipe stress analyses, site conditions, and the purpose of the thrust pipe. For preliminary design, using an approximation of 100 times the product pipe diameter, a 36 in. (915 mm) outer diameter product pipe would have a bend radius of 100 × 36 in. = 3,600 ft (1,100 m). The typical metric approximation for the preliminary design radius is the diameter (in meters) times 1,000, plus 10%.

5.2.3 Launch and Reception Tangents

During design, consideration should be given to providing a launch tangent at least long enough to allow the MTBM to pass through the entry seal. Where possible, the launch tangent should be long enough to allow the MTBM and the initial few joints of the thrust pipe to advance in a straight line. The longer tangent allows the contractor to obtain initial optical survey control, without requiring personnel entry into the thrust pipe to validate MTBM survey readings. The launch tangent is typically 200 to 500 ft long (60 to 150 m) but can vary depending on the site conditions and crossing geometry needed. A reception tangent is not required but, if possible, should be considered to avoid higher induced stresses within the overbend connecting the DSPT installation with the mainline pipeline. If adequate tangent lengths are not possible, provisions should be considered to properly tie in the mainline pipeline.

5.2.4 Minimum Depth Requirements

The depth of cover requirements for a DSPT installation are a function of

- Anticipated ground conditions;
- Anticipated depth of scour (if applicable);
- Presence of preferential flow pathways for slurry and/or lubrication flow and inadvertent drilling fluid surface release risk;
- Permit requirements for separation distance from existing utilities, levees, railroads, highways, and/or other structures; and
- Acceptable magnitudes of systemic settlement for a given installation.

Typical minimum installation depths are on the order of three times the outer diameter of the thrust pipe. The ground overlying the DSPT installation must be able to resist the required lubrication and slurry pressures to avoid excessive loss of fluids to the surrounding environment, surface, and crossing feature(s). The ability of the ground to resist the required pressures is a function of the strength of the anticipated subsurface conditions (Chapter 4, Section 4.3). The depth of cover will need to consider any additional requirements imposed by the owner of the crossing feature (e.g., railroad requirements and levee district requirements).

For applications where ground cover is less than three times the outer diameter and where geotechnical and construction risks are well defined and acceptable, special measures (such as ground improvement) and carefully balanced advance and excavation rates may be used to mitigate risk and allow for the DSPT installation. Evaluation and calculation of the potential settlement and heave risk should be assessed during design, particularly in the vicinity of sensitive structures such as pavements, utilities, and railroads. This is addressed in Section 5.3.3.

5.2.5 External Fluid Pressure and Maximum Installation Depth Limitations

The MTBM is followed by one or more trailing cans that are connected together (Chapter 2, Section 2.3.2). Seals are placed within the joints of each section to isolate and prevent intrusion of groundwater, slurry, or lubricating fluid into the inner workings of the MTBM. MTBMs have an internal ring seal that is rated for an external pressure of approximately 45 psi (300 kPa or 3 bar). On MTBM diameters of 36 in. (915 mm), and larger, the ring seals can be upgraded to withstand a pressure of approximately 60 psi (400 kPa or 4 bar). MTBMs of 48 in. (1,219 mm) in diameter, and larger, are capable of achieving pressure ratings of up to 70 psi (500 kPa or 5 bar). The fluid pressure limitation for an MTBM must consider the external pressure acting on the outside of the MTBM derived from both groundwater (including confined aquifers, if applicable) and the lubrication fluid column as well as the face pressure. If designers are concerned about the anticipated external pressure for a given installation, they should contact the MTBM manufacturer to determine if an appropriate MTBM is available to suit their project needs. In addition, consideration should be given to potential high internal pressure in the slurry system for DSPT crossings with significant elevation differential.

Depending on the unit weight of lubricating fluid and the groundwater conditions, the standard 3-bar seal is typically capable of withstanding depths of roughly 80 to 100 ft (24 to 30 m). A 4-bar seal is typically capable of accomodating depths of approximately 110 to 130 ft (33 to 40 m).

5.3 ENGINEERING DESIGN CONSIDERATIONS

5.3.1 Evaluation of Thrust Forces

During DSPT installation, the thrust force required to advance the MTBM through the design profile consists of three main components: (1)

required face pressure to support and excavate the encountered material, (2) the frictional resistance acting on the outside of the MTBM and thrust pipe, and (3) the weight/pulling force of the pipe string leading into the pipe thruster. The anticipated thrust forces must be evaluated and understood to design the thrust pipe, pipe thrusting system, and foundation to withstand the installation loads and stresses. The project specifications should require contractor submittals to confirm adequate capacities.

The criteria for estimating thrust forces should consider the subsurface conditions specific to the site as provided in the geotechnical report. The thrust force calculations also require the selection of appropriate geotechnical and fluid properties based on engineering judgment and the results of the geotechnical investigation.

Engineers should use sound engineering judgment when evaluating methods to estimate the total thrust forces. The thrust forces should be estimated using industry-recognized rational methods; however, a few methods have been thoroughly evaluated to estimate the thrust forces through curved sections. Engineers may consult conventional microtunnel methods to estimate thrust forces (Bennett and Cording 2000, Staheli 2006, Burnett and Gelinas 2018), although these methods were developed for straight microtunnel sections and they typically overestimate thrust loads for DSPT. Until field-verified, equation-based methods to estimate thrust forces for DSPT can be validated, Pruiksma et al. (2012) and Rabiei et al. (2018) may be considered. For simplified estimates of maximum thrust forces, Pfeff (2013), Robison and Hotz (2014), and Sparks and Hotz (2019) provide guidance based on empirical data.

Similar to jacking forces in microtunneling, thrust forces with DSPT consist of three primary components: pipe string force, face pressure, and frictional resistance. The total thrust force can be expressed as follows:

$$F_T = F_S + F_P + \sum F_R \tag{5-1}$$

where

F_T = Total thrust force,
F_S = Pipe string force,
F_P = Face pressure force, and
F_R = Frictional resistance (acting in the direction of the pipe axis).

5.3.1.1 Pipe String Force. The pipe string force (F_S) is typically very small when compared to the face pressure and frictional resistance forces but should be evaluated if the pipe string is very long or has a significant slope. If the pipe string is relatively long and slopes down to the launch pit, it is possible for the pipe string force to be negative (Pruiksma et al. 2012) and for the thrust pipe to require restraint.

5.3.1.2 Face Pressure Force. The face pressure force (F_P) component may be estimated as the product of the cross-sectional area of the MTBM head and the sum of the groundwater pressure and effective earth pressures (Burnett and Gelinas 2018).

In practice, the measured face pressure is typically slightly greater than the combination of groundwater and active earth pressure acting on the cutter face. If the applied face pressure falls significantly below the sum of groundwater and active earth pressure in unstable ground conditions, overmining with associated ground losses at the face may occur, potentially resulting in ground surface settlements. If the measured face pressure significantly exceeds the combination of groundwater and passive earth pressure, ground surface heave may occur.

Face pressure is typically a relatively small component of the overall thrusting force. The face pressure measured during construction is typically driven by the penetration rate and force applied by the thruster. The measured face pressure should be compared to the design face pressure to evaluate the risk of overmining or heave.

5.3.1.3 Friction Force. The frictional resistance force (F_R) component is the force on the pipe as the thrust pipe moves through the design tunnel profile. Methods presented by Pruiksma et al. (2012) and Rabiei et al. (2018) may be considered to estimate the frictional forces.

The effective normal stress acting on the pipe may be evaluated with arching theory and is proportional to the thrust pipe's outer diameter and ground conditions. Calculations of effective normal stress using total overburden significantly overestimate the normal stresses acting on the MTBM and the pipe, except in relatively shallow installations (i.e., less than 2 diameters of earth cover).

For drives in very stiff to hard clay and stable rock, effective normal stresses acting on the thrust pipe may be very small. The frictional component in these cases may be more accurately estimated as the buoyant weight of the thrust pipe multiplied by an appropriate effective friction factor between the pipe and the ground.

5.3.1.4 Factors That Influence Thrust Forces. A variety of factors can affect the thrust forces. The following are some examples of factors that can affect the thrust forces:

- Ground conditions,
- Thrust pipe buoyancy,
- Effective normal stress acting on the thrust pipe and the MTBM,
- Effective friction factor along the thrust pipe,
- Condition and use of engineered drilling fluid and slurry,
- Condition and use of lubricating fluid,

- Overcut dimension,
- Face pressures caused by earth and hydrostatic pressures,
- Steering of the MTBM and the pipe string, and
- Pipe curvature through the design curved sections.

Some of these factors are not explicitly considered when calculating thrusting forces using available methods. The key to reasonable accuracy is the selection of appropriate ground properties on the basis of a thorough geotechnical investigation and a fundamental understanding of the geotechnical characteristics that influence or control behavior.

Operational factors can exert significant influence on the estimation of thrusting forces but are difficult to quantify. For example, if excessive slurry pressure is used, fines may be forced around the cutter wheel and into the overcut, which will significantly increase thrust forces. As the MTBM advances, cuttings created by the MTBM cutterhead can be forced into the overcut, significantly increasing thrusting forces (and torque) if the lubricating fluid pressure is not high enough and the cutter wheel does not efficiently clean the invert. If lubrication is not used or is used ineffectively, thrust forces can also significantly increase. These and other operational factors are typically not within the engineer's control. Therefore, the engineer making the thrust force calculations typically assumes proper equipment setup and good workmanship. All assumptions should be clearly stated when these and all other design calculations are made.

5.3.1.5 Buoyancy Assessment and Control. Recent research indicates that the buoyancy of the thrust pipe can have a significant effect on the maximum thrust force (Sparks and Hotz 2019). A thrust pipe that is nearly neutrally buoyant can exhibit a much lower thrust force than a highly buoyant thrust pipe for an equivalent drive length and subsurface material. As a result, the buoyancy of the thrust pipe should be evaluated. Where possible and feasible, the wall thickness of the thrust pipe should be optimized to achieve near neutral buoyancy and reduce the maximum thrust force. The weight of the umbilical lines will also affect the overall buoyancy of the thrust pipe, as discussed in Chapter 2, Section 2.3.6.

5.3.2 Hydraulic Fracture Evaluation

Although DSPT uses a closed slurry system to transport cuttings and a separate lubricating system to lubricate the annulus, hydraulic fracture of the ground can occur. Hydraulic fracture will occur if the fluid pressure at the face, or the pressure of the lubrication fluid, exceeds the confining strength of the surrounding ground. As a result, the risk of hydraulic fracture, potentially leading to inadvertent return of fluids to the surface, should be evaluated.

The method commonly used in the trenchless industry to evaluate the potential for hydraulic fracture is derived from cavity expansion theory. The procedures used to evaluate the potential for slurry loss through hydraulic fracturing of soil are based primarily on research completed by Delft Geotechnics, as initially discussed in Appendix B of United States Army Corps of Engineers (USACE) Report CPAR-GL-98 (Staheli et al. 1998), and more recently in USACE Engineer Manual EM-1110-2-2902 (USACE 2020). The cavity expansion model is used to estimate the maximum effective pressure or formation limit pressure in the drilled hole or tunnel before plastic deformation of the surrounding soil occurs. The estimation of the formation limit pressure considers the total and effective overburden stresses as well as shear strengths of the soil. In rock, fluid loss risk is typically controlled by the structure of the rock, discontinuities, and other factors, and cavity expansion models are not the appropriate method to assess hydraulic fracture and inadvertent drilling fluid return risk. Currently, there is no accepted standard for the evaluation of fluid loss in rock.

Along with the formation limit pressure, both slurry and lubricating fluid pressures must be estimated. As stated previously, cuttings produced during the tunneling operation are conveyed to the ground surface through the slurry system. The pressure in the excavation chamber is adjusted to stabilize the tunnel face (face pressure) while advancing the MTBM by counterbalancing the earth and groundwater pressures. To facilitate face stability, a slight overpressure (relative to the surrounding, in situ earth and groundwater pressure) is typically applied and ranges between 1.5 and 6.0 psi (10 to 40 kPa or 0.1 to 0.4 bar).

The lubricating fluid is injected into the annular space through a lubrication ring located between the MTBM and the pipe string. This fluid is injected at a slightly overpressure to fill the annulus with fluid. If the lubricating fluid is continuous through the overcut annulus, the pressure associated with the lubricating fluid can be estimated by multiplying the height of the fluid column by the density of the lubricating fluid. The height of the column of lubricating fluid must consider the aboveground tank elevation, if used.

5.3.3 Overcut and Settlement

The radial overcut for DSPT installations is typically larger than 1 in. (25 mm) but can be more than 6 in. (150 mm), depending on the diameters of the MTBM and the thrust pipe. The larger (than typical microtunneling) overcut is used to lower thrust forces and accommodate curved pipeline installation geometry. The overcut should be as small as possible, while still reducing the anticipated thrust loads to those that the pipe thruster can provide and that the thrust pipe can withstand.

With larger overcuts, significant settlement can result (especially for shallow installations) and should be evaluated for potential impacts to surface and subsurface infrastructure. Because existing utilities, roadways, railroads, foundations, structures, and other facilities must be protected against damage from ground movements resulting from DSPT installations, the potential settlement should be appropriately evaluated and mitigated to reduce the risk to those existing facilities. The steps to evaluate and mitigate the risks of settlement include the following:

- Identify facilities or features at risk for settlement;
- Establish recommended maximum allowable settlements for critical features in conjunction with facility owners and other stakeholders;
- Evaluate potential settlements at each feature;
- Compare estimated settlements against recommended maximum allowable settlements for each feature and develop mitigation measures to reduce the effects of settlement, if necessary;
- Develop a settlement instrumentation and monitoring plan for the project and require that the plan be implemented during construction;
- Coordinate with permitting agencies and owners of critical features or facilities to identify and formulate mitigation steps to avoid damage that may occur as a result of DSPT construction;
- If the evaluation results indicate high risks of damage, it may be necessary to move or deepen the DSPT alignment and/or relocate facilities or protect them in place.

Systemic settlement associated with DSPT are those associated with closure of the annulus between the overcut and the pipe. Additional systemic settlements associated with overmining (overexcavation) of the face may also occur, but these are typically minor. The risks of systemic ground movement associated with DSPT should be evaluated using the approach refined for trenchless construction methods by Wallin et al. (2008).

The approach approximates the settlement as a trough above a DSPT installation in the shape of an inverted normal probability distribution curve. Representative soil properties must be selected using judgment combined with the results of the geotechnical investigation. Systemic settlements are proportional to annular space and pipe diameter, and they are inversely proportional to soil stiffness, depth of earth cover, and offset distance from centerline of the pipeline alignment. The first step in evaluating anticipated settlements is to calculate the annular space as the difference between the volume of the excavated tunnel and the pipe volume. The next step is to estimate the proportional amount of the annular space that is likely to contribute to the settlement trough volume. For stable rock, the proportion of annular space that contributes to the settlement trough volume may be close to zero; for very loose fill and debris, it may approach 100%. For DSPT installations, the volume of annulus that can collapse and contribute to the settlement trough volume may also be reduced by the volume of bentonite-based slurry and lubrication fluid.

5.3.4 Engineered Drilling Fluid, Slurry, and Lubricating Fluid Design

Engineered drilling fluid, slurry, and lubricating fluids are essential components of DSPT design and construction. The two major components of both engineered drilling fluid and lubricating fluids are water and naturally occurring bentonite clay. Other additives could include pH modifiers, polymers, clay inhibitors, detergents, or other additives to enhance the fluid's desirable engineering properties, enhance the ground behavior, reduce formational loss, aid in the carrying capacity of solids, reduce equipment wear, and promote lubrication. As a consideration of overall environmental impact, DSPT typically requires significantly less water and generates lower disposal volumes, relative to HDD.

The fluids used in a DSPT installation should be engineered to the specific ground conditions expected to be encountered by the MTBM and modified as appropriate during construction. Testing during construction should be performed to monitor the rheologic properties of the slurry and lubricating fluids. Typically, testing is performed both on the drilling fluid being sent to the cutterhead and on the slurry returned to the separation plant. Additional information about drilling fluid, slurry, and lubricating fluids and testing is provided in the ASCE 36-15 Microtunneling Standard (ASCE 2015).

5.3.4.1 Engineered Drilling Fluid and Slurry. The primary purpose of the engineered drilling fluid is to support the excavation face and to mix with and transport spoils to the separation plant at the ground surface. The engineered drilling fluid consists primarily of water and, depending on the anticipated ground conditions, it often contains bentonite, polymers, or other additives to enhance its desirable engineering properties. If only water is considered, overexcavation may result in cohesionless ground (Boyce et al. 2011, Boyce and Joyal 2018).

The slurry system is a closed loop whereby the engineered drilling fluid is transported from the ground surface through the feed lines to the excavation face. The engineered drilling fluid mixes with the excavated spoils and becomes slurry. The slurry is then transported back to the ground surface through the slurry return lines.

The engineered drilling fluid should be designed based on the anticipated ground conditions and to accomplish the following:

- Counterbalance the earth and groundwater pressures at the face of the MTBM,

- Mix the engineered drilling fluid with excavated material and transport the slurry to the surface for separation,
- Efficiently transport the slurry to the separation plant without excessive pressures or velocity,
- Reduce the potential for overexcavation, and
- Allow for efficient excavation.

The composition of an engineered drilling fluid must be considered in advance, and the rheologic properties are tested regularly to evaluate its performance. Slurry management should always be considered a key element to successful DSPT installation.

5.3.4.2 Slurry Recycling System Considerations. Similar to the design of the engineered drilling fluid, the slurry recycling system should be designed for the anticipated subsurface conditions based on the results of the subsurface evaluation program (Chapter 4, Section 4.3).

Depending on the subsurface conditions, the mechanical components of the slurry recycling system could include the following:

- Settlement basins or tanks,
- Shakers with appropriate vibrating screens,
- Desanding cones,
- Desilting cones,
- Centrifuges,
- Polymer flocculation or mixing units, and
- Vertical clarifiers.

Because the subsurface profile of each DSPT installation is unique, these components should be combined based on the subsurface condition to best separate the spoils from the slurry. The configuration of the recycling system should be designed by a drilling fluid technician (often referred to colloquially as a mud engineer) who is familiar with the anticipated subsurface conditions and the properties of the drilling fluids.

Managing the slurry recycling plant requires planning, customizing of equipment, monitoring, and revising as the excavated ground changes. The planning starts by reviewing the project geotechnical reporting and evaluating the subsurface conditions. When the ground conditions are understood, the contractor should select the proper components and additives. These items help determine operating and disposal costs. The contractor should review the slurry recycling plan with their material suppliers and adjust the plan as needed.

5.3.4.3 Lubricating Fluid. Lubricating fluid is pumped into the annular space between the excavated tunnel created by the gauge cut and the body of the MTBM to reduce the friction between the thrust pipe and the surrounding ground. Lubricating fluid is not intended to carry cuttings but should be designed to reduce friction and fluid loss. As a result, it may use different additives, including polymers, to accomplish this. Specifically, the lubricating fluid should be designed to accomplish the following:

- Counterbalance the earth and groundwater pressure along the length of the drive,
- Create a filter cake to prevent the loss of fluid to the ground formation, and
- Provide lubricity and reduce the friction between the thrust pipe and the excavated tunnel.

The lubricating fluid is typically introduced through the launch seal and from one or more special lubrication rings located between the MTBM machine can and the trailing section. Because the thrust pipe is often the product pipe, lubricating fluid is not typically introduced to the annulus through ports in the thrusted pipe as is common in microtunneling.

5.3.5 Launch Area Considerations

The MTBM can be launched from the ground surface or from a shallow shaft/pit. If launched from the ground surface, consideration should be given to the use of a conductor casing to provide fluid containment, support of adverse ground, and a controlled launch for the MTBM.

5.3.5.1 Launch Workspace Considerations. Because nearly all the support equipment for a DSPT installation is located on the launch side, the launch workspace must be sized to accommodate the equipment and supplies. The ground surface in the launch workspace is typically armored with matting to facilitate all-weather construction. Appropriate erosion control is required to manage stormwater runoff. Although the shape of the launch workspace can be flexible, the required area is on the order of 200 ft (60 m) wide by 200 ft (60 m) long. A launch workspace is shown in Figure 5-2.

5.3.5.2 Launch Ground Control and Other Geotechnical Considerations. The MTBM may be launched from the ground surface or a shaft/pit. Regardless of whether it is launched from the surface or a shaft/pit, it is typically launched through a rubber-gasketed launch seal (as described in Chapter 2, Section 2.3.7) attached to a head wall or conductor casing.

Use of a launch shaft/pit can decrease the amount of equipment necessary to support the overbend section of the pipe string. The launch shaft pit excavation may be supported by sheet piles or other shoring installed on three sides, with the fourth side open and sloped to allow for

Figure 5-2. Athabasca River crossing launch workspace.
Source: Courtesy of Michels Corporation.

the passage of the pipe string into the pipe thruster (as shown in Figure 5-2). The main benefits of using a launch shaft include a shortened installation length and reduced pipe-handling equipment needs associated with the overbend section of the pipe string. Where a headwall is used, it is beneficial to provide a depth of cover at the launch location that is a minimum of two times the product pipe diameter to provide resistance to counter the required slurry and lubrication pressures and to facilitate ground arching. A minimum width of approximately 30 ft (10 m) is needed to assemble and stage the pipe thruster within the shaft/pit. The length of the base is dependent on the depth of the shaft but should be greater than about 50 ft (15 m).

The design of a shoring system should be completed by a professional engineer to provide adequate temporary shoring capacity. If the presence of groundwater is expected above the shaft floor, dewatering should be considered to maintain dry working conditions and reduce the shoring requirements.

In some cases, an oversized steel casing is used to mount the launch seal. The inside of the steel casing can be lined with a polyvinyl chloride or high-density polyethylene liner to reduce the risk of damage to the thrust pipe. The conductor casing is typically removed on completion of the installation. If the steel casing is driven into the ground, the design line and grade can be difficult to maintain. As a result, consideration should be given to the possibility that the launch angle and alignment could differ from the design, and a plan should be implemented to address design or construction practices to address the issue.

An additional geotechnical consideration is the foundation system for the pipe thruster. The thrust loads required to advance the pipe must be transferred from the pipe thruster to the ground without excessive deformation or movement of the pipe thruster. This is often done with a deep foundation, such as driven piles or drilled piers, depending on the geotechnical conditions and contractor preferences.

5.3.5.3 Pipe Stringing Area. Although DSPT is capable of thrusting short sections of pipe, pipe stringing areas should be as long as practical to reduce the number of welds and the associated downtime and cost during welding. The pipe stringing area should extend in-line with the pipeline alignment and be horizontally straight, where practical. Horizontal curvature in the stringing area should be limited to those portions after the pipe breakover section and should be within the acceptable bending capabilities of the pipe. If horizontal curvature of the pipe string is required, mitigation will be required to keep the thrust pipe from rotating during installation. The actual length of the pipe string should be based on site conditions, taking into consideration the following:

- Thrust forces,
- Topography,
- Subsurface conditions,
- Available workspace,
- Location of the slurry recycling system,
- Site access,
- Site grade, and
- Construction schedule.

For planning purposes, the stringing area width should be at least 50 ft (15 m) plus 20 ft (6 m) for each additional string.

To facilitate an internal survey without a personnel entry into the thrust pipe, it is often advantageous to launch with a relatively short pipe section, less than 500 ft (150 m) in length, so the survey can be completed while the pipe string is still in the launch tangent. Longer sections are then welded to the pipe string to complete the installation.

5.3.6 Reception Area Considerations

The primary purpose of the reception area is to recover the MTBM and trailing sections and make the necessary tie-in connections to complete the pipeline crossing. For DSPT installations, only a few pieces of equipment are needed on the reception side. The area can also be much smaller than

Figure 5-3. Typical reception area.
Source: Courtesy of GeoEngineers.

the launch area, and typically, a 50 to 100 ft wide (15 to 30 m) by 100 to 150 ft (30 to 45 m) long area is needed.

The reception shaft/pit for a DSPT installation is typically not a shored excavation unless the recovery depth or other site constraints require shoring. If conditions do not require shoring, the reception pit side walls should be sloped in accordance with safe excavation practices. If shoring is required, a minimum shored length is necessary to retrieve the MTBM and trailing cans. If groundwater is relatively high, dewatering of the reception shaft/pit area could be required to safely recover the MTBM and other equipment. An example of a DSPT reception area is shown in Figure 5-3.

5.3.7 Recommended Installation Tolerances

Direct steerable pipe thrusting is capable of installation tolerances similar to those typically required for microtunneling [i.e., 6% and 3% of the MTBM diameter for line and grade, respectively (ASTM 36-15)]. However, DSPT installations often incorporate larger tolerances for the installation of force mains, high-pressure pipelines, or casings. Allowing larger installation tolerances reduces the number of personnel entries required to complete internal surveys to calibrate the internal line and grade measurement systems.

Installation tolerances should be evaluated on a project-specific basis. Consideration should be given to the size of the thrust pipe because personnel entry for surveying in pipes of 36 in. (915 mm) and smaller is not practical. The cost of installation will be higher when maintaining tighter tolerances because additional survey work is required.

REFERENCES

ASCE. 2015. *Standard design and construction guidelines for microtunneling no. 36*. New York: ASCE.

Bennett, D., and E. J. Cording. 2000. "Jacking loads and ground deformations associated with microtunneling." In *Proc., North American Society for Trenchless Technology No-Dig Show*, 17–36. Cleveland, OH: NASTT.

Boyce, G. M., and N. Joyal. 2018. "Microtunneling with water only can cause overexcavation." In *Proc., North American Society for Trenchless Technology No-Dig 2018*. Paper TM1-T5-03. Cleveland, OH: NASTT.

Boyce, G. M., M. Wolski, R. Zavitz, and C. Camp. 2011. "Chemistry and physics behind microtunnel slurries." In *Proc., North American Society for Trenchless Technology No-Dig 2011*. Paper A-2-01. Cleveland, OH: NASTT.

Burnett, A., and M. Gelinas. 2018. "Revised method of estimating microtunnel jacking forces." In *Proc., North American Society for Trenchless Technology No-Dig Show*. Paper TA-T5-01. Cleveland, OH: NASTT.

Pfeff, D. 2013. *Overview of thrust forces needed on realized Direct Pipe projects*. Schwanau, Germany: Herrenknecht.

Pruiksma, J. P., D. Pfeff, and H. M. G. Kruse. 2012. "The calculation of the thrust force of pipeline installation using the Direct Pipe Method." In *Proc., Pipeline Technology Conf.* https://www.pipeline-conference.com/abstracts/calculation-thrust-force-pipeline-installation-using-direct-pipe-method.

Rabiei, M., K. P. Chang, M. Gelinas, and A. Neale. 2018. "Analysis and design of pipes installed via Direct Pipe Technology." In *Proc., North American Society for Trenchless Technology No-Dig Show*. Paper WM-T4-02. Cleveland, OH: NASTT.

Robison, J., and R. Hotz. 2014. "Direct Pipe—Estimated and actual installation load analysis for 20 crossings." In *Proc., North American Society for Trenchless Technology No-Dig Show*. Paper TM2-T5-04. Cleveland, OH: NASTT.

Sparks, A., and R. Hotz. 2019. "Direct Pipe—Installation force evaluations of recent Direct Pipe installations." In *Proc., ASCE UESI Pipelines Conf.* Reston, VA: ASCE.

Staheli, K. 2006. "Jacking force prediction: An interface friction approach based on pipe surface roughness." Ph.D. thesis, Georgia Institute of Technology, School of Civil and Environmental Engineering.

Staheli, K., D. Bennett, H. W. O'Donnell, and T. J. Hurley. 1998. "Installation of pipelines beneath levees using Horizontal Directional Drilling." Final Rep. CPAR-GL-98-1. Washington, DC: US Army Corps of Engineers, Waterways Experiment Station.

USACE (US Army Corps of Engineers). 2020. "Conduits, pipes, culverts associated with dams and levee systems." *Engineer manual EM-1110-2-2902*. Washington, DC: USACE.

Wallin, M., K. Wallin, and D. Bennett. 2008. "Analysis and mtigation of settlement risks of new trenchless installations." In *Proc., North American Society for Trenchless Technology No-Dig Show*. Paper C-4-03-1. Cleveland, OH: NASTT.

CHAPTER 6
STEEL PIPE DESIGN

6.1 INTRODUCTION

This chapter covers the following topics as they relate to the direct steerable pipe thrusting (DSPT) method of trenchless steel pipe installation.

- Installation-related axial, bending, hoop, torsional, and combined stresses;
- Installation-related pipe buckling evaluation and considerations;
- Allowable thrust force calculation;
- Operational stresses; and
- Pipe-coating considerations.

6.2 INSTALLATION-RELATED STRESSES

As the steel pipe is thrusted through the ground, stresses are induced in the pipe that must be assessed to avoid overstressing the pipe and yielding of the steel. If yielding occurs, it could result in an unsafe condition during installation and/or during subsequent pressurization of the installed pipe during its operating life. The reader is encouraged to review the steel design codes and other documents referenced for all detailed mathematical equations. They are summarized in this chapter for brevity.

6.2.1 Axial Stress

To calculate the design factor of safety for axial compressive stress, the applied axial stress (f_a) is calculated from the estimated maximum load by dividing the load by the cross-sectional area of the thrust pipe:

$$f_a = \frac{F_t}{A} \tag{6-1}$$

where F_t is the applied load (total thrust force as discussed in Chapter 5), and A is the cross-sectional area of the pipe.

The allowable axial stress (F_a) is then calculated using the equations in American Petroleum Institute (API) Recommended Practice 2A-WSD (API 2014), Section 3.2.2, depending on the pipe diameter-to-thickness (D/t) ratio, the steel yield strength (F_y), the unbraced length, and Young's modulus of elasticity for steel. To check whether the design is acceptable, the applied compressive stress must be less than the allowable stress. The calculations in the referenced API code consider local buckling effects; however, column buckling considerations often control the allowable axial stress, as described in Section 6.3.

6.2.2 Bending Stress

To calculate the design factor of safety for bending stress, the applied stress (f_b) is calculated from the following equation derived from beam mechanics as presented in the PRCI design guide (Hair 2015):

$$f_b = \frac{ED}{24R} \tag{6-2}$$

where

E = Steel pipe Young's modulus of elasticity,
D = Outside pipe diameter (in inches), and
R = Radius of pipe curvature (in feet).

The allowable bending stress (F_b) is calculated using the equations in API (2014), Section 3.2.3; depending on the pipe D/t ratio and the steel design strength. To check whether the design is acceptable, the applied bending stress must be less than the allowable stress.

6.2.3 Hoop Stress

Two loading conditions should be evaluated: (1) the hoop stress imposed by earth and groundwater pressure, and (2) the hoop stress imposed by the pipe thruster clamp to calculate the design factor of safety for hoop stress.

The allowable hoop stress (F_h) is calculated using the equations in API (2014), Section 3.2.5, which considers both elastic and inelastic hoop buckling stress. To check whether the design is acceptable, the anticipated hoop stress must simply be less than the allowable stress.

6.2.3.1 Ground, Drilling Fluid, and Groundwater—Hoop Stress. During installation, the ground, drilling fluid, and water surrounding the pipe exert stress on the outside of the thrust pipe that is not counteracted by internal pressure. The applied stress (f_h) is calculated from the estimated external and internal pressures by the following equation from the PRCI design guide (Hair 1995):

$$f_h = \frac{\Delta p D}{2t} \tag{6-3}$$

where

Δp = Difference in pressure inside the pipe (assumed to be at atmospheric pressure) and outside the pipe from earth, groundwater, and drilling fluid;
D = Outside pipe diameter; and
t = Pipe wall thickness.

6.2.3.2 Thruster Clamp—Loading Considerations. As described in Chapter 2, the pipe thruster clamp consists of several pads that are typically hydraulically actuated, curved, rubber-lined, and articulated—providing the gripping force necessary to hold and push the pipe.

For the Herrenknecht pipe thrusters (HK500PT and HK750PT), this gripping of the pipe imposes an external hoop stress of up to approximately 0.5 ksi (3.5 N/mm^2) (Pfeff 2013). However, the stress imposed by the pads is not continuous owing to the gaps between pads, which results in nonuniform stress concentrations. The clamping pressure has the potential to damage seam welds and propagate any linear defects found within the weld joint. Accordingly, it is preferable to use a pipe with longitudinal, straight seams rather than a spiral-welded pipe and to stagger thruster strokes to avoid clamping on or within typically one pipe diameter of a circumferential girth weld. Other pipe thruster manufacturers have different gripping systems and impose different stresses that should be evaluated.

6.2.4 Torsional Stresses

The rotating cutting wheel at the face of the microtunnel boring machine (MTBM) produces a torsional stress on the thrust pipe. The maximum torsional shear stress (F_{vt}) is calculated using equations in API (2014), Section 3.2.4b, and this value is compared to the allowable torsional shear stress which is defined as 40% of the steel yield strength (F_y). This value should be compared to the selected MTBM's maximum torque.

A typical MTBM used in DSPT is capable of producing torque on the order of 35 to 150 kips-ft (50 to 200 kN·m), depending on the MTBM

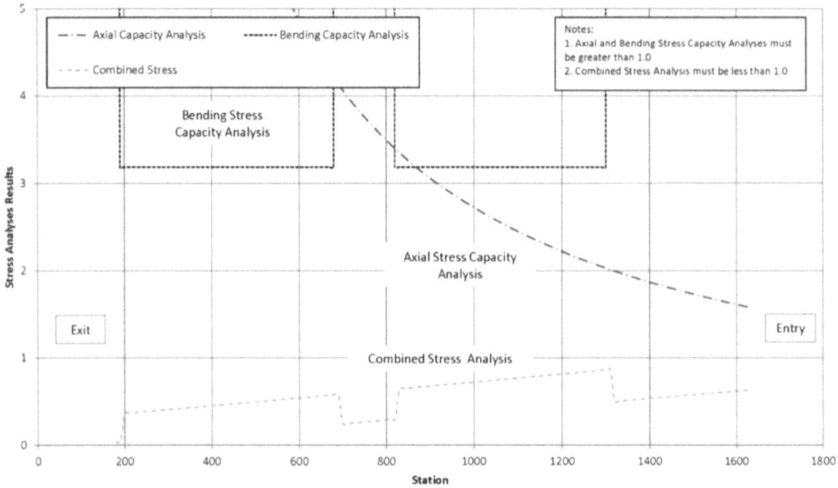

Figure 6-1. Example capacity analysis graph.
Source: Courtesy of GeoEngineers.

diameter. At these torque values, torsional stress is not typically a design criterion that requires special consideration for longitudinal welded pipe. Torsion may require special consideration if the thrust pipe is spiral-welded.

6.2.5 Combined Stresses

Combined loading and stress conditions occur during DSPT installation. API (2014), Section 3.3, provides the equations that must be satisfied (typically results of less than or equal to 1.0) to confirm adequate pipe capacity in combined stress conditions. A graphical example of the results of combined stress calculations along the length of the thrust pipe is shown in Figure 6-1.

6.3 PIPE BUCKLING ANALYSIS AND CONSIDERATIONS

Buckling is defined as the occurrence of a lateral deflection of a long, slender structural member subjected to an axial compressive load. For an ideal column, the critical load (P_{cr}) at which buckling will occur is calculated by using the standard equation developed by Euler buckling theory:

$$P_{cr} = \frac{\pi^2 EI}{(KL)^2} \tag{6-4}$$

where

E = Steel pipe Young's modulus of elasticity,
I = Pipe section moment of inertia,
K = Effective length factor, and
L = Length of the thrust pipe.

The engineer should consider the risk of lateral movement and buckling of the installation pipe during DSPT installation.

During DSPT installation, the cutter wheel (and sometimes the MTBM itself) is larger than the diameter of the thrust pipe. In many applications, this has resulted in overcuts of up to several inches radially. The resulting overcut annulus is filled with bentonite lubrication fluid, groundwater, and spoils from the excavation. The degree to which this annulus is filled and the degree to which the material in the annulus prevents lateral movement of the thrust pipe during advancement are poorly understood. Accordingly, for the purposes of this manual of practice, column buckling is assumed to occur. If we use conventional Euler buckling theory as described, the relatively long unbraced lengths of thrust pipe (relative to a building column, for instance) typically predict buckling at axial loads that are much lower than are necessary to install the pipe.

If the thrust pipe buckles inelastically during DSPT installation, yielding of the steel and a significant loss of strength are likely, which would negatively affect the integrity and performance of the pipe over its operational life. This is a particular concern if the thrust pipe is to be used in a high-pressure application.

The American Institute of Steel Construction (AISC) provides guidance on flexural buckling in the ANSI/AISC 360-10 standard (AISC 2016), Chapter E. American Institute of Steel Construction presents a limit state of flexural buckling defined by a *critical stress* F_{cr}. The equations are used to determine the critical stress curve for two regions of the slenderness ratio, Kl/r, and present the critical stress curve for inelastic flexural buckling and the critical stress curve for elastic flexural buckling. The critical buckling stress at the dividing point between elastic and inelastic behavior is calculated by

$$F_{cr} = 0.877\ F_e = 0.877 \frac{\pi^2 E}{\left(4.71\sqrt{E/F_y}\right)^2} = 0.39 F_y$$

where

F_{cr} = Critical buckling stress,
F_e = Elastic buckling stress,
F_y = Steel minimum yield stress, and
E = Steel pipe Young's modulus of elasticity.

The AISC flexural buckling equations, therefore, limit the elastic flexural buckling stress to 39% of the steel yield stress. The equations also imply that portions of the thrust pipe will experience inelastic deformation when the column buckles at a stress level higher than this value. For additional discussion, see Robison and Chen (2017).

If inelastic buckling occurs, yielding of the steel pipe is possible with a resulting potential reduction of capacity during operation. Yielding is especially critical to avoid for a thrust pipe that will be subjected to high product pressures (natural gas or oil pipelines, for example). The end restraint, boundary condition, residual stress, and geometric imperfection of the thrust pipe are difficult to define precisely. Accordingly, the allowable axial stress should be limited to 39% of the steel yield strength to limit any buckling to the elastic range and to avoid yielding of the pipe.

6.4 ALLOWABLE THRUST FORCE CALCULATION

The procedure for estimating thrust forces is presented in Chapter 5. The design engineer should factor the thrust force estimate in accordance with their knowledge (or lack thereof) of the geotechnical conditions at the site, the capacity of the pipe thruster, their previous experience, and other considerations. Use of this factored value, along with the desired pipe specification, will serve as the basis to estimate pipe stresses in the design geometry. Using the equations provided and through the stress analysis described, the engineer should confirm that the pipe specification and design geometry are appropriate for the factored load or provide recommendations to change the pipe specification and/or design geometry to accommodate it. After confirming the adequacy of the thrust pipe to tolerate the factored load, the engineer should run the calculations in an iterative manner to define a limiting state or maximum allowable thrust force for a given pipe specification in the design geometry. This limiting force should be included in the design documents as a not-to-exceed requirement during construction. There are two methods of performing the stress calculations: a maximum combined stress scenario and a distributed stress scenario.

6.4.1 Maximum Combined Stress

In the maximum combined stress scenario, the engineer considers that the maximum thrust force (and axial stress) occurs at the location along the pipe of maximum bending stress. This results in a theoretical maximum load and stress combination on the pipe. This method gives no allowance for distribution of axial load along the thrust pipe through skin friction and can result in limiting the combined stress calculation results, particularly if the crossing is long, with a long entry tangent.

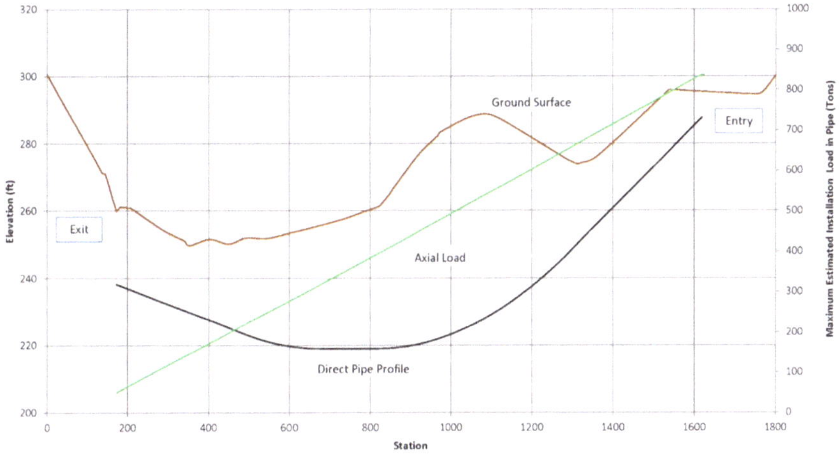

Figure 6-2. Example load distribution in a thrust pipe just prior to MTBM exit. Source: Courtesy of GeoEngineers.

6.4.2 Distributed Axial Stress

In the distributed axial stress scenario, the engineer develops a theoretical load distribution on the installation pipe, considering both the face pressure and the skin friction along the length of the thrust pipe as the MTBM approaches exit (the theoretical maximum load condition), an example of which is shown in Figure 6-2.

This load distribution is converted to theoretical maximum axial stresses along the length of the thrust pipe. Bending and hoop stresses are also calculated along the length of the pipe. An example of the results of these calculations is shown in Figure 6-3.

Continuing with the design process, the installation stresses discussed and presented in Figure 6-3 are compared to allowable stresses (reduced in accordance with the allowable stress and buckling discussion presented) and combined stresses are calculated through the length of the proposed crossing. The allowable stresses divided by the estimated installation stresses are presented as the pipe "capacity" and must be greater than or equal to 1.0. As discussed, the combined stress analysis must be less than 1.0. An example of the capacity and combined stress analysis is shown in Figure 6-1.

6.5 OPERATING STRESSES

For thrust pipes subjected to pressure operations, the operating stresses should be calculated considering test and transient pressures in

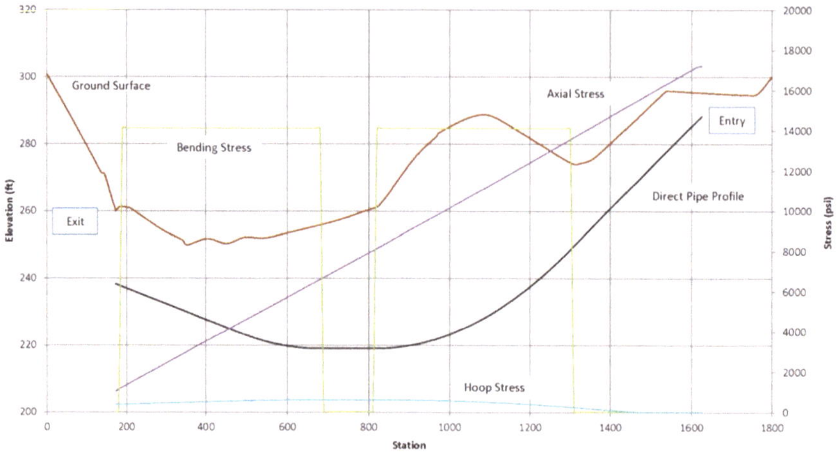

Figure 6-3. Example stress distribution in a thrust pipe just prior to exit. Source: Courtesy of GeoEngineers.

addition to the normal operating conditions. For a steel pipeline subject to internal pressure, the PRCI HDD design guide provides clear instructions for the calculations of bending, hoop, longitudinal, shear, thermal, biaxial, and combined operational stresses (Hair 1995). These calculations are available from other references and are therefore not presented herein for brevity.

6.6 STEEL PIPE COATING AND LINING

Most steel pipelines are coated externally. A pipeline coating provides a layer on the outside of the pipe that protects the pipeline from the damaging effects of corrosion. If coatings are not required, the pipe wall thickness can be increased to serve as sacrificial steel to account for corrosion, or other corrosion protection methods may be employed. Depending on owner requirements and pipe use, linings may be applied to the inside of the pipe. Coating and lining used in a DSPT application must be flexible to accommodate installation bending. Corrosion protection of the pipe is beyond the scope of this MOP and an expert should be consulted; the following sections address DSPT installation considerations related to external pipe coating.

As with other trenchless installation methods, the risk of abrasion to the pipe coating from the subsurface conditions encountered during installation should be considered. In addition, for DSPT installations, the shear stress on the coating to pipe bond developed from the pipe thruster clamp must be evaluated.

6.6.1 Coating Types

Many coating types are available in the industry. Some of the more commonly used coatings are described subsequently.

6.6.1.1 Fusion-Bonded Epoxy. Fusion-bonded epoxy (FBE) is an epoxy-based powder coating that is typically applied in the mill and is used to protect steel pipes used in pipeline construction. FBE coatings are thermoset polymer coatings. The name fusion-bonded epoxy comes from the resin cross-linking and the application method. Resin and hardener in the dry FBE powder remain neutral under normal storage conditions. At application temperatures in the range of 355 to 480 °F (180 to 250 °C), the powder melts and is transformed into a liquid that flows over the steel surface on which it is applied. This process is known as fusion bonding as the coating reaches a thermoset. Application of additional heat will not melt the coating nor change the thermoset of the coating. As a pipeline coating, FBE provides good dimensional stability.

6.6.1.2 Abrasion-Resistant Over-Coating. Abrasion-resistant over-coating (ARO) is used to provide physical protection to FBE pipeline coatings. ARO is typically a mill-applied overcoating designed to protect the FBE basecoat from damage during trenchless installation. It offers protection to coated pipes installed in abrasive ground.

A typical ARO system consists of an initial primer layer of 8 to 16 mils (0.2 to 0.4 mm) of epoxy covered with 20 to 35 mils (0.5 to 0.9 mm) of a higher-density epoxy. Liquid, 100-percent-solid epoxy ARO coatings can be spray-applied at a thickness of up to 60 mils (1.5 mm). Field application may also be possible, reference Section 6.6.3.

6.6.1.3 Polymer Concrete. Polymer concrete (sometimes referred to as liquid ARO) is sometimes used as a protection layer for the FBE coating on steel pipes. Polymer concrete is a mixture of fine aggregate and epoxy binder that provides an abrasion and impact resistant coating. Polymer concrete can be sprayed onto FBE-coated pipes at a thickness range of 20 to 125 mils (0.5 to 3 mm). Its suitability for DSPT applications must be verified on a case-by-case basis.

6.6.2 Coating Stress

When a coated pipe is to be installed with DSPT, the suitability and integrity of a particular coating must be verified beforehand because the pipe thruster grabs or clamps the pipe around its perimeter on the outside and hence applies a clamping force to both the pipe and the coating. The pipe thruster must generate enough friction to transfer its full thrust force to the coating without damaging it, and subsequently, the coating must transfer this force to the steel pipe underneath.

Although the forces—both clamping and pushing/pulling—are considerable, the induced stresses are governed by the contact area of the clamping device. Consideration needs to be given by the coating designer to the nonuniform stress distribution of the clamp, given the areas of no load between pads. Because different manufacturers use different coating designs, coating stresses need to be addressed on a case-by-case basis. For example, the maximum stress imposed by a Herrenknecht pipe thruster clamp is approximately 145 psi (1 N/mm^2) for shear stress and around 435 to 500 psi (3 to 3.5 N/mm^2) for compression stress.

6.6.3 Coating Application

It is recommended all pipe coatings be factory applied. When compared to field application, factory application provides better production efficiency, a controlled environment, better access for application and quality control, and the ability to apply complex coating systems that cannot be applied in the field. Field-applied coatings are required at the joints where the pipes are welded in the field and at areas requiring repair. The pipe thruster stroke should be staggered to avoid clamping in any areas of field-applied coating. As a general rule, clamping should be avoided within one pipe diameter of a joint.

The field-applied joint coating needs to provide a smooth transition to the pipe's factory-applied coating. Overlapping coating areas with protruding edges or seams that are not flush bear a risk of being damaged during the installation process.

6.6.4 Other Materials

Direct steerable pipe thrusting is currently applicable only for steel thrust pipe. Pipe materials other than steel have been installed as carrier pipes within the steel casing installed by the DSPT process. Materials such as polyvinyl chloride or high-density polyethylene are used commonly as carrier or product pipes.

REFERENCES

AISC (American Institute of Steel Construction). 2016. *Specifications for structural steel buildings*. Chicago: AISC.

API (American Petroleum Institute). 2014. *Planning, designing and constructing fixed offshore platforms-working stress design*. 22nd ed. Washington, DC: API.

Hair, J. 2015. *Installation of pipelines by horizontal directional drilling an engineering design guide*. Houston: Pipeline Research Council International.

Pfeff, D. 2013. *Method statement—Direct Pipe®*. Schwanau, Germany: Herrenknecht.

Robison, J., and C. Chen. 2017. "Considering buckling—Direct Pipe® engineering design." In *Proc., NASTT No-Dig Conf.*, Paper No. TM2T5-02. Cleveland, OH: NASTT.

CHAPTER 7
CONTRACT DOCUMENTS

7.1 INTRODUCTION

Contract documents are a combination of legally binding written documents prepared by the owner and engineer that define the roles, responsibilities, and work specified under the construction contract by the contractor. These typically include the following: instructions for bidding, commercial terms of agreement between the owner and contractor, general and special conditions that outline the scope of work requirements to be followed, construction specifications that present the project-specific technical requirements of the project, and geotechnical data. In addition, for some projects, the ground conditions are contractually defined in a Geotechnical Baseline Report (GBR) (ASCE 2007).

Contract documents for direct steerable pipe thrusting (DSPT) projects share much in common with other trenchless pipeline and utility projects in that they should be structured to clearly present the legal and commercial terms, scope of work, technical aspects, responsibilities of each party, ultimate expectations of the project, and the basis of payment. It is important to note that DSPT is a specialty construction technique with unique technical considerations. This chapter focuses on the contract documents for DSPT projects. Contract forms, delivery methods, and potential legal issues such as differing site condition (DSC) clauses and alternative dispute resolution (ADR) are also briefly addressed.

7.2 CONTRACT FORMS

The contract form dictates who is responsible for the project design and how the contractor will be paid for the work. The most common contract forms for DSPT projects can be separated into three basic categories:

- Lump sum,
- Time and materials, and
- Turnkey (design-build).

A lump sum or fixed-price contract is one in which the owner agrees to pay an independent contractor a negotiated price for the work regardless of any increases or decreases in the contractor's costs to produce the work (Black and Garner 1999). The contractor determines the means and methods for accomplishing the project within the framework of the contract documents. Prices for DSPT installations under lump sum contracts may be structured based on a total project cost or as a cost per foot of accepted pipe installed. Lump sum contracts are often considered *no hole, no pay*, which is a term originating from the oil and gas industry, meaning payment will not be made to the contractor if the installation is not completed in accordance with the owner's design, drawings, and specifications. The drawings and specifications are supporting and supplementary information structured to control the final characteristics and quality of the construction (i.e., the specifications are tailored to the contractor's performance requirements) and not the contractor's means and methods that are used to construct the crossing.

A time-and-material (T&M) contract is one in which the contractor is paid an agreed-upon fee amount per day, or some other unit of time, for providing supervision, equipment, and labor in accordance with the contract documents. In this contract type, the owner shares the responsibility of co-managing and coordinating project responsibilities and progression because of its principle that payment is based on the passage of time regardless of progress made in the installation.

A turnkey or design-build contract has much in common with a lump sum contract, with the primary difference being that the contractor also assumes the responsibility for designing the crossing. In an Engineering, Procurement, Construction contract, a form of turnkey contracting, the contractor also assumes the responsibility of procuring the materials.

Contract form preference varies from industry to industry and even within specific industries by different owners. Lump sum contracts are favored in the oil and gas industry by both owners and contractors, whereas T&M and design-build contracts are often used by other private and municipal owners.

7.2.1 General Conditions

The general conditions of a construction contract define the requirements, responsibilities, and relationship between the owner and contractor of a project. The contents of the general conditions vary by owner and contract form (see aforementioned section). Common topics covered in the general conditions may include, but are not limited to

- Project-specific scope of work;
- List of bid items;
- Legal framework for enforcing the contract, including definitions, indemnity, measurement, and payment;
- Insurance requirements;
- Material requirements;
- List of owner-supplied materials;
- Disposal sites secured by the owner;
- Schedule constraints or limitations to working hours;
- Permit requirements;
- Site-specific sound limitations;
- Contractual means of dealing with obstructions;
- Change order procedures;
- Dispute resolution procedures; and
- Liquidated damages for failure to complete the project on time or nonperformance.

7.2.2 Special or Supplementary Conditions

Special or supplementary conditions may be necessary for a project depending on contract form and owner preference. These conditions may include landowner requirements and regulatory agency requirements. These conditions often define the documents that were or can be relied on during bid preparation. The special provisions typically include the DSC clause and an order of precedence for the contract documents, especially if a GBR was prepared as part of the contract documents.

7.2.3 Technical Specifications

A technical specification is a document that defines a minimum set of requirements that must be met, minimum requirements for material conformance, and minimum performance requirements for acceptance of each element of work. The technical specifications for a project typically include separate sections for each major item of work. The technical specifications should clearly describe each major item, relevant standards, quality assurance and quality control, submittal requirements, work products, and execution.

Technical specifications can be broadly classified as either performance specifications or prescriptive specifications. Performance specifications establish the requirements for the product to be constructed, with means and methods at the discretion and the responsibility of the contractor.

Prescriptive specifications establish the requirements for the resulting product and select (or constrain) the means and methods required to be used by the contractor to construct the specified product. The designer should be aware that prescriptive specifications may create an implied warranty by the owner that if the contractor uses the means and methods required, with generally accepted industry good practices, the resulting product will meet the requirements of the contract.

In practice, most technical specifications are a combination of both approaches. The specification may be primarily a performance specification but may contain some prescriptive elements to limit the contractor's range of options. In general, the completed DSPT projects in North America have predominantly used performance-based rather than prescriptive-based specifications.

Technical specifications for a DSPT project include many of the same specifications that would be included in the contract documents for other trenchless installation methods such as horizontal directional drilling; however, some specifications for other installation methods will need to be altered because of the unique nature of DSPT projects. As a general guideline, technical specifications for a DSPT project should include, but not be limited to

- DSPT operations,
- Pipe material,
- Clearing and grading,
- Hauling and stringing,
- Manual arc welding of steel piping,
- Field application and repair of pipe coatings,
- Pipe handling and stringing,
- Field testing, and
- Cleanup and site restoration.

To maintain consistency across the project specifications, the sections should be written in the same general format. Because many of the technical specifications are standard specifications common to most pipeline and utility projects, this chapter does not address all specifications that may be included in a DSPT contract. Regardless of format or approach, all specifications should be checked for project-specific consistency.

7.2.3.1 Technical Specification for Direct Steerable Pipe Thrusting Installations. The technical specifications covering DSPT installations should convey the following information at a minimum:

- Minimum experience and qualifications of the contractor's key personnel,
- Product or carrier pipe specification and coating requirements (including field joints),
- General DSPT equipment requirements,
- Required preconstruction submittals,
- Daily reporting requirements,
- Horizontal and vertical alignment tolerances,
- Minimum acceptable radius of curvature,
- Maximum allowable thrust forces,
- Maximum allowable ground settlement,
- Carrier pipe or casing support and handling requirements,
- Launch shaft/pit performance requirements,
- Acceptance testing requirements,
- Inadvertent returns monitoring and response requirements,
- Spoil disposal requirements, and
- As-built documentation.

When a product or carrier pipe is installed inside the thrust pipe (casing), the material, coating, and testing requirements of the product or carrier pipe may be addressed in the DSPT specification or under a separate specification section.

7.2.4 Plans

In addition to the technical specifications, the contract documents should contain plans pertinent to the crossing signed and sealed by the engineer of record responsible for the design. This section assumes that the owner is responsible for the DSPT design and plans. However, in some instances (e.g., design build contracts), development of the plans is the responsibility of the contractor. At a minimum, the plans should complement the technical specification by providing a clear presentation of the crossing design as well as the results of topographic, hydrographic, and geotechnical surveys. The plans will be used by the contractor to produce a working profile, which aids in downhole navigation, making accurate measurements essential. In most cases, the contractor performing the work also requires the electronic survey data and drawings for the purpose of site layout.

As a general guideline, the plan and profile drawings should convey the following information:

- Ground surface and bathymetric contours within 100 ft (30 m) of the crossing alignment;
- Survey control points;
- DSPT launch and reception shaft/pit locations referenced to a known coordinate system (State Plane, UTM);
- Horizontal and true length of the crossing;
- Geometry of the DSPT design, including station and elevation of the points of curvature/points of tangency, and radius of curvature;
- Pipe diameter(s), wall-thickness, grade, and coating requirements;
- All known above and below ground utilities and structures, both active and abandoned;
- Installation tolerances (horizontal, vertical, and radius);
- Maximum thrust force;
- Exploratory boring locations and subsurface data;
- Temporary workspace and permanent right-of-way; and
- Sensitive environmental features.

Appropriate directives should advise the contractor about the requirement to carry out the exercise of utility locates and gather other information necessary for the contractor to execute the work.

7.3 GEOTECHNICAL RISK ALLOCATION

In general, the most significant risks associated with trenchless projects, including DSPT projects, are delays, damage, or losses that result from unanticipated, poorly understood, or misinterpreted subsurface conditions. Owners and engineers have varying philosophies and practices regarding how subsurface information and interpretation are presented and disseminated to bidders. All geotechnical data collected during various phases of the project should be compiled, provided to bidders, and become a part of the contract documents, and this is typically accomplished through the use of a Geotechnical Data Report (GDR). Contractual interpretations of the subsurface conditions may be established and presented in a GBR, as discussed in Section 7.3.2. Preliminary and detailed geotechnical investigations and reporting are described in Chapter 4.

7.3.1 Differing Site Conditions

A DSC is defined as an unknown physical condition encountered at a site that differs materially from the reasonably anticipated conditions (ASCE 2007). There are varying approaches to address DSCs.

Many owners, notably in the oil and gas pipeline industry, have traditionally expected their contractors to account for subsurface risks in a fixed price (lump sum) contract with the expectation that it is the contractors' responsibility to bid projects with enough contingency included to offset these risks. This belief was also typical in other areas of heavy civil construction and can be summed up in the expression *You bid it—you build it*. In this scenario, the contractor must absorb any additional costs that may result from unanticipated subsurface conditions such as encountering hard rock where soft ground conditions are expected. The owner, on the other hand, should expect and absorb inflated bid prices that have large contingencies to account for unknown ground conditions.

This contracting philosophy may be satisfactory in relatively routine installations where subsurface conditions are uniform and predictable. However, the owner should expect that the contractor will submit a request for additional compensation and schedule relief (if applicable) if the construction process encounters unexpected subsurface conditions that result in a material change to schedule or cost. This is typically referred to as a DSC claim.

A method of resolving disputes between the owner and the contractor needs to be included in the contract documents. It is recommended that the contract, at a minimum, include a DSC clause to allow the contractor to pursue compensation for extra costs without being forced into a breach of contract situation. As an example, US government contracts include a DSC clause (Code of Federal Regulations title 48 52.236-2) that has been tested in court and found to be enforceable.

7.3.2 Use of a GBR in the Contract

For a DSC clause to achieve its objective of allocating risk associated with adverse or changed ground conditions, there must be a clear definition or understanding of what the subsurface conditions are or may be, and how they will respond relative to the specified construction method. As discussed in Chapter 4, a GBR is the contractual document that can provide this definition. It functions as an interpretation of the GDR and should supersede it in the contract document hierarchy, when used. The purpose of the GBR is to establish baseline ground conditions and ground behaviors on which the bidders can base their bids. The baselines must be specific to the intended construction method (i.e., DSPT) and should reflect available geotechnical data, as well as engineering judgment and owner input. During construction, the ground conditions and behaviors encountered are then monitored relative to the established baselines. The baselines used must be clear, observable, and measurable. If the conditions and behaviors encountered are within the established baseline parameters, the contractor is responsible for the conditions. If the conditions or behavior are outside of the bounds or definition of the baseline condition, a DSC may exist. The objective of a GBR is to balance risk between the owner and the contractor.

The statement given subsequently, taken from ASCE's *GeotechnicalBaseline Reports for Construction: Suggested Guidelines* (originally attributed to the US National Committee on Tunneling Technology), describes the desired intended result when providing a GBR (ASCE 2007)

> ...if all bidders can base their estimates on a well-defined set of site conditions with assurance that equitable reimbursement will be made when changed conditions are encountered, the owner will receive the lowest reasonable bids with a minimum of contingency for unknowns.

A detailed discussion of techniques for preparing a GBR, including suggested format, organization, and content, is presented in ASCE's *Geotechnical Baseline Reports for Construction: Suggested Guidelines* (ASCE 2007). Although the concept of the GBR was developed for tunneling projects, the methods described are applicable to a wide range of subsurface construction activities, including DSPT installations.

The benefits of a GBR are realized only if the GBR is concise, well-written, comprehensive, appropriate to the technology, measurable, and the limitations and implications are understood by the owner and contractor. In addition, robust construction monitoring is necessary to document encountered site conditions. As with all geotechnical data and reporting, philosophies vary widely among owners regarding the use of GBRs. GBRs have yet to see significant use in the oil and gas industry; however, they are commonly used on public works projects.

7.4 ALTERNATIVE DISPUTE RESOLUTION

Disputes that cannot be resolved in a timely matter may cause project and payment delays, thereby increasing costs for all parties. To help alleviate this situation, contracts can be structured to require intervention byADR measures for unresolved disputes (DRBF 2007). For underground projects, the disputes are often complicated and intermeshed with complex engineering and machinery that requires specialized expertise that a Dispute Resolution Board (DRB) brings to the ADR process. Underground public sector projects use a DRB as an established and accepted method to resolve disputes. The DRB is typically composed of one expert each selected by the contractor and owner and a third board member jointly selected by the owner's and the contractor's representatives. It is common to have these board members selected early and to meet quarterly to keep them abreast of the work during the course of construction.

The DRB process has proven to be a successful method for resolving claims and disputes since it was first employed in 1975. Simply put, a DRB is a group (typically three members) of impartial and knowledgeable professionals formed at the beginning of a project who are familiar with the details of the project, encourage dispute avoidance, and assist in the resolution of disputes for the duration of the project (DRBF 2007). In general, DRB decisions are nonbinding, unless stipulated in the contract, so the parties are free to pursue legal recourse. However, the ADR process often reveals the strength and weaknesses of each party's disputes, an important factor in deciding to pursue further legal avenues.

A DRB typically provides significant advantages over litigation or arbitration. DRB action is often quick, impartial, and carried out by technically strong professionals who understand the issues, with differences resolved while the project is ongoing. The cost of the DRB is typically less than an adversarial drawn-out court process.

7.5 CONTRACTOR PREQUALIFICATIONS

Prequalifying a DSPT contractor or subcontractor should be considered for projects with identified high-risk elements. Prequalification elements can include financial security or bonding capacity, technical expertise with DSPT technology and high-risk construction elements, experience in the geographic area and local ground conditions, resumes of key personnel, and experience in anticipated site conditions. The benefit of prequalification is to limit the bidders to those likely to produce a successful project. However, technical prequalification is not always necessary or possible because of public bidding laws.

REFERENCES

ASCE. 2007. *Geotechnical baseline reports for construction: suggested guidelines.* ASCE, Technical Committee for the Underground Technology Research Council. New York: ASCE.

Black, H. C., and B. A. Garner, eds. 1999. *Black's law dictionary.* 7th ed. St. Paul, MN: West Group.

DRBF (Dispute Resolution Board Foundation). 2007. *DRBF practices and procedures.* Section 1. Seattle: Dispute Resolution Board Foundation.

US Government Information. "United States Code of Federal Regulations title 48 part 52.236-2 'Differing site conditions.'" www.govinfo.gov.

CHAPTER 8
CONSTRUCTION CONSIDERATIONS

8.1 INTRODUCTION

This chapter summarizes the direct steerable pipe thrusting (DSPT) construction process from preconstruction activities through as-built documentation. The following topics are discussed as they relate to the construction phase of DSPT installations:

- Safety,
- Preconstruction activities and submittals,
- Staging and layout,
- Launch preparations,
- Machine operations,
- Contingency measures, and
- Construction documentation and as-built submittals.

8.2 SAFETY

8.2.1 Site Safety

As with all construction projects, DSPT construction requires a site-specific construction safety plan. Because each project is unique and site conditions vary, the safety plan should be developed for each phase of the DSPT construction.

8.2.2 Project-Specific Safety Program

The construction safety program should be developed following award of the construction contract. The safety plan should address contract requirements, be site-specific, and should be developed by the contractor with input from the owner's representative and general contractor. The safety plan should be comprehensive and address all construction activities associated with DSPT installation (Carpenter et al. 2018).

8.2.3 Confined Space Safety Plan

Most of the DSPT project experience in North America to date has been for the oil and gas industry, where owners and contractors have classified the thrust pipe as a confined space and followed regulations and standards for confined space entry into the pipe during construction. The following section addresses requirements for confined space entry.

8.2.3.1 Confined Space Entry and Requirements. There are two types of confined space entry into the thrust pipe. Scheduled entry includes entering the thrust pipe to complete planned optical survey control or to complete planned maintenance on the microtunnel boring machine (MTBM) or other internal equipment. Unscheduled entry includes entering the thrust pipe to make repairs to damaged or malfunctioning equipment or to complete additional survey control if the location of the MTBM is questioned.

Details of the scheduled confined space entry are typically described in the DSPT execution plan. In general, only as many entries into the confined space as necessary should be scheduled. When possible, entry into the confined space should combine purposes to reduce the number of entries and the amount of downtime.

The number of scheduled confined space entries for survey control depends on the length of the DSPT installation, the complexity of the design, and required installation tolerances. Typically, at least one confined space entry should be scheduled after the profile enters the launch curve to verify whether the internal survey controls are accurate. Additional confined space entries for optical survey control may be required to verify whether the alignment and profile are within the allowable tolerances set forth for the project.

Unscheduled confined space entries are often required to make repairs, perform maintenance, or complete additional survey control if internal survey control indicates that the alignment and profile are out of tolerance. Although the timing of these entries is not necessarily predictable, the DSPT contractor should plan for at least one unscheduled confined space entry.

8.2.3.2 Confined Space Entry Plan. The safety requirements during construction for pipe entry are typically more rigorous and detailed than for other construction or trenchless projects. Entry into the thrust pipe is allowed only for personnel trained and certified for the confined space entry requirements associated with the DSPT installation. A detailed confined space entry plan needs to be developed by confined space professionals experienced with DSPT projects and in compliance with government regulations.

The confined space entry procedures need to include at a minimum the following:

- Onsite Confined Space Rescue Team (CSRT),
- Confined space entry permit,
- Authorized attendant maintaining constant communication with personnel in the confined space,
- Air quality inside the thrust pipe monitored and evaluated continuously while personnel are inside the confined space,
- Continual ventilation of compressed air inside confined space, and
- Access to emergency respirators inside the confined space.

8.2.3.3 Confined Space Rescue Teams. The CSRT is required when personnel are inside the thrust pipe. The CSRT should assist in the development of the confined space rescue plan and provide knowledge of all aspects of the planned rescue procedures. The confined space emergency rescue plan needs to include the following:

- Project information,
- Confined space classification,
- Rescue response procedures, and
- Rescue communication protocols.

8.2.4 Preconstruction Safety Orientation

Although the DSPT installation procedure is similar to aspects of microtunneling and horizontal directional drilling (HDD), the workspace layout and construction activities are different and require all personnel working on the site to be familiar with and aware of site-specific hazards. Prior to beginning construction activities for a DSPT installation, all parties who will be on site during construction should complete a safety orientation specific to DSPT. This safety orientation is often led by the DSPT contractor but could also be led by the owner representative or general contractor. During the preconstruction safety orientation, all parties should give their commitment to abide by the project safety plan.

8.2.5 Owner-Specific Safety Requirements

In addition to the contractor-developed safety program, some owners have additional safety guidelines and procedures in place that augment or supersede those generated by the contractors. These measures need to be incorporated into the project safety plan, reviewed during the preconstruction safety orientation, and followed during construction.

8.2.6 Job Hazard Analysis

The construction practices of DSPT are similar to both microtunneling and HDD, and many of the hazards associated with construction are similar. However, there are enough differences in the construction practices that a site-specific job hazard analysis (JHA) should be completed for every construction process employed during DSPT installation. The JHA should be completed by the DSPT contractor and be accepted by the owner prior to beginning construction activities.

8.2.7 Weather Safety

A site-specific and seasonal-specific weather safety plan should be a part of the overall DSPT construction safety plan. The plan should include how to protect personnel from severe weather, excessive heat, or extreme cold weather.

8.3 PRECONSTRUCTION ACTIVITIES AND SUBMITTALS

8.3.1 Contractor Qualifications

It is a good practice to either shortlist prequalified contractors for bidding and execution of a DSPT project or use a qualification-based selection process. The level of DSPT experience required for the contractor depends on the challenges, complexity, and risks involved in a particular project. When evaluating contractor qualifications, the experience of the proposed key field personnel should be considered. Key field personnel include the superintendent, machine operator, electrician, mechanic, drilling fluid specialist, and surveyor.

8.3.2 DSPT Execution Plan

It is common practice to submit a preliminary DSPT execution plan (or project approach) during the bidding phase of the project. After the contract is awarded, a final execution plan is submitted by the DSPT contractor that incorporates changes based on additional information available. These changes may include DSPT design modifications, staging and workspace modifications, pipe fabrication and handling support, and any services provided by third parties or the owner. Typical sections included in a final execution plan are described subsequently.

8.3.2.1 List of Key Personnel. A list of key personnel, their responsibilities, and their contact information along with an organizational chart should be included in the final execution plan. The organization chart should identify the engineer of record, owner's representatives, and other third-party contractors or consultants.

8.3.2.2 Schedule. The contractor should establish a baseline schedule with their best estimate of expected construction duration. The schedule needs to list the intended daily work hours and weekly workdays for each phase of the project. The schedule should also establish significant and critical path tasks and milestones throughout the duration of the project with durations and anticipated completion dates. The schedule needs to be updated regularly to capture any changes resulting from site conditions and durations of intermediate tasks and milestones.

8.3.2.3 Material and Equipment Specifications. The execution plan (or required submittals) should identify the equipment and materials that the DSPT contractor intends to use for execution of the project. The identified equipment should include the MTBM with the proposed cutter wheel type and tooling configuration, pipe thruster(s), and a drilling fluid separation system. Typically, an MTBM corresponding to the outside diameter of the thrust pipe is used. In some cases, the contractor may propose the use of an oversized MTBM. When oversizing is proposed, systemic settlement should be checked against the design engineer's criteria.

The pipe thruster needs to be sized for the anticipated maximum thrust forces (Chapter 5, Section 5.3). If two pipe thrusters working in series are proposed to provide greater thrust capacity, the maximum thrust force on the pipe calculated by the design engineer cannot be exceeded (Chapter 6, Section 6.3).

The drilling fluid separation system needs to be sized and configured to manage the slurry resulting from the anticipated maximum advance rates. Inadequate drilling fluid cleaning capacity will cause a reduction in productivity.

8.3.2.4 Drilling Fluid Plan. An important consideration for DSPT construction operations is the drilling fluid (Chapter 2, Section 2.2.1). The contractor may use water or a mixture of water and bentonite, sometimes with other additives based on anticipated subsurface conditions. A drilling fluid plan should provide a list of the potential drilling fluid mixes and their properties that the contractor intends to employ based on available subsurface information. Safety data sheets (SDSs) for bentonite and other additives should be identified within the execution plan. The SDS must be present on site when these materials are on site.

8.3.2.5 Solid and Fluid Disposal Plan. The drilling fluid separation plant (Chapter 2, Section 2.3.8) needs to separate a significant quantity of solids from the slurry. The solid and fluid disposal plan should be provided in the execution plan and it should describe the details of solid and drilling fluid transportation and disposal. Unless a preapproved disposal site has been provided by the owner, the execution plan should include the location of a contractor-proposed disposal site. The disposal plan should include the following:

- Documentation of acceptance of both solid and liquid wastes by the disposal facility;
- Requirements for any environmental or other tests on disposal samples;
- Required maximum moisture content for disposed solids (note, this can be difficult to estimate prior to construction);
- Contractor's plan to use a drying agent to dry cuttings (if applicable). If used, the SDS for the drying agent should be included.
- Site storage/detention capacity for wastes; and
- Methodology to handle hazardous/contaminated material, if anticipated or encountered.

8.3.2.6 Workspace Layout Drawings. Workspace layout drawings should depict the location of equipment planned for DSPT operations. The workspace layout drawings should confirm that adequate space is available for the execution of the project. Equipment shown should include the pipe thruster, control cabin, generators, and drilling fluid separation plant. Location of storage items such as tool van, spare part containers, and bentonite pallets should also be shown. Mobile equipment such as excavators, forklifts, and cranes should be shown for reference to verify adequate moving space. Access and parking spaces for personnel vehicles should be clearly marked. The contractor should evaluate the workspace for adequate turnaround space for large vehicles such as trailer trucks, semitrucks, vacuum trucks, and dump trucks that require regular ingress to and egress from the job site. The thrust pipe string(s) should be shown on the workspace layout drawing.

8.3.2.7 Launch Details. The design drawings will not typically specify the launch details of a DSPT installation. The decision to launch from a shored entry shaft or a surface launch is normally left to the discretion of the DSPT contractor. The contractor should include the launch details in their execution plan.

If surface launching, the plan should include the following:

- Diameter and length of the launch casing;
- Details of the launch seal and connection of seal to casing;

- Wall thickness and length of the launch casing and whether it includes a poly vinyl chloride or high-density polyethylene insert;
- Depth of the cover at the end of the casing;
- Method of installing the casing;
- Dimensions, depth, and configuration of shallow excavation, if any; and
- Location of the pipe thruster and support foundation.

If launching through a shored launch shaft, the execution plan should include the following:

- Details of excavation sequencing and dewatering (if needed),
- Dimensions of the shaft,
- Maximum depth of the shaft,
- Shoring details including bracing,
- Details of the launch seal and connection of seal to shoring,
- Tail ditch details, and
- Location of the pipe thruster and support foundation.

If the DSPT design changes to accommodate the launch methodology, a plan and profile drawing needs to be submitted and reviewed for conformance with the design intent and standards established by the owner's engineering team.

8.4 LAYOUT AND STAGING

In DSPT construction, most of the equipment is staged at the launch workspace until the crossing is completed and the MTBM is retrieved. Typically, MTBM retrieval takes place at the surface. However, in certain situations, a reception shaft or pit with additional equipment and workspace will be required. Different aspects of site layout and staging are discussed subsequently.

8.4.1 Preconstruction Survey

Preconstruction survey for DSPT consists of verifying available workspace and laying out or staking the major design components. Preconstruction survey typically includes a detailed location verification of shoring, foundation, and other elements and the establishment of benchmarks during the equipment setup process. Once the field entry point is established, which may vary from the planned entry point, the contractor should reevaluate and adjust the thruster setup to accommodate workspace development or other adjustments and constraints. In some cases, the final entry point may need to be adjusted after the thruster and other equipment is set up.

If the contractor intends to use temporary structures in lieu of equipment to support the thrust pipe during tunneling operation, the locations of such structures should be reevaluated and staked based on the established final entry point and thruster orientation. An external, fixed reference point should also be established to allow for survey checks during initial setup and construction. Once the pipe thruster and the MTBM are set up, the preconstruction survey concludes with an evaluation of the MTBM location, elevation, and heading relative to the design alignment. The MTBM survey information is entered into the navigation system. A high degree of accuracy is required as the steering of the MTBM is based on this final input data.

8.4.2 Site Preparation

A stable, all-weather worksite is necessary for the support of DSPT equipment. The launch side of the DSPT installation should be prepared to provide adequate access for the DSPT equipment, thrust pipe string, and support equipment. This preparation is often accomplished by placing timber mats, but other options may be appropriate, particularly in urban settings. The contractor should consider the following key components while performing the site preparation work:

- Adequate space at the entry point area for the entry shaft/pit or launch casing and pipe thruster with the control cabin;
- Workspace for the slurry system with adequate room for the fluid separation and bentonite mixing equipment, as well as hauling equipment ingress and egress. Siting of the fluid systems should also consider containment of the slurry in the event of equipment malfunction and other environmental considerations.
- Work area for thrust pipe fabrication and stringing to allow for welding and handling of multiple pipe strings, if required;
- Work area for slurry feed and discharge lines, lubrication lines, and other umbilical components from the tail end of the thrust pipe string to the slurry separation, bentonite system, and control cabin; Typically, these components travel over the ground surface as the operation progresses. These surfaces should be maintained as clean as practicable and the lines protected from damage.
- Last, adequate area for construction vehicle loading and offloading, turnaround, egress, and parking.

8.4.3 Equipment Layout

DSPT contractors have individual preferences on equipment setup based on their experience, their actual equipment configuration, and site-specific conditions. Some of the key equipment requirements and considerations are as follows:

- Control cabin is placed adjacent, if practical, to the pipe thruster such that the operator has a clear view of the pipe thruster during operation.
- If possible, the drilling fluid cleaning system is typically staged at the middle of the thrust pipe stringing area. This setup is so that the maximum distance from the drilling fluid cleaning system to the tail end of the installation pipe at any particular stage of the tunneling operation is kept to a minimum.
- When a separate power-pack module is used for the DSPT installation, most contractors prefer to place it on a trailer with its dedicated generator at the end of the pipe string to reduce umbilical length and size. Adequate space is needed for the trailer-mounted power-pack module to travel down the stringing area as the pipe string is being installed. Alternatively, the power-pack module may be located at the middle of the stringing area or it can be integrated with the control cabin. A typical layout is shown in Figure 2-14.
- Sensor for the hydrostatic water level (HWL), which is part of the navigation and guidance system, needs to be located above the maximum height of pipe overbend.

8.4.4 Launch Configuration

8.4.4.1 Shoring and Shaft Construction. As discussed in Chapter 5, Section 5.2, DSPT design geometry may require an entry point below the ground surface to achieve desirable entry/exit angles, bottom tangent elevations, and annular hydrostatic pressures. The contractor may choose to use an entry pit or shaft as opposed to a ground surface launch because of the following advantages:

- Increased ground confining pressures with depth, reducing the risk of inadvertent drilling fluid surface release;
- Reduction in the overbend height of the thrust pipe, resulting in a reduction of required pipe handling and support equipment; and
- Elimination of an oversized launch casing, which is typically required for a surface launch.

Sheet pile shoring is typically used for entry shaft construction. Depending on the depth and requirements of safety regulations at the project location, the entry shaft should be designed in accordance with applicable regulations and owner requirements.

The dimensions of the entry pit depend on the contractor's equipment, the anchoring or bracing system used, and the contractor's preference on additional workspace in the entry pit. Depending on the depth of the groundwater table, the contractor may also be required to perform dewatering or use a watertight shoring system.

8.4.4.2 Surface Launch. A surface launch eliminates or reduces the requirement of a launch shaft. A conductor casing is installed at the planned entry angle to support the MTBM, provide initial ground stabilization, and contain drilling and lubrication fluids. It is important that adequate depth of casing be provided so that the risk of surface release of drilling fluid is reduced. If site conditions allow, fill can be placed above the casing to increase the effective overburden pressure and reduce the risk of undesired drilling fluid surface release. Refer to Chapter 3, Section 3.4.1 and Figure 3-8.

8.4.4.3 Launch Seal Installation. The purpose of the launch seal is to keep the lubrication fluid in the overcut annulus throughout the DSPT process. The launch seal is especially important when the exit point or any other part of the DSPT profile is at a higher elevation than the entry point. The launch seal should be designed to resist additional groundwater pressure for the case where the groundwater phreatic surface is above the elevation of the launch seal. In an entry shaft launch, the seal is installed on the face of the shoring by cutting out a portion of the wall and mounting the launch seal at the opening; the excavation support should be designed to accommodate this penetration. For a surface launch, the seal is installed on the entry conductor casing.

8.4.5 Pipe Thruster Foundation

8.4.5.1 Design and Reaction Forces. It is common practice to design the thruster foundation for achieving the maximum capacity of the pipe thruster. This may result in a thruster foundation with a significantly higher capacity than needed for the actual DSPT installation. At a minimum, the pipe thruster foundation should be designed for the anticipated thrust forces, as discussed in Chapter 5, Section 5.3, including an appropriate factor of safety.

Figures 8-1 and 8-2 show a typical schematic of the pipe thruster and the reaction forces. Equations (8-1) to (8-3) calculate the reaction forces depicted in the free body diagrams.

$$R_A = (F \times \cos\alpha \times H + F \times \sin\alpha \times L2)/(L1 + L2) \tag{8-1}$$

$$R_B = (F \times \cos\alpha \times H - F \times \sin\alpha \times L1)/(L1 + L2) \tag{8-2}$$

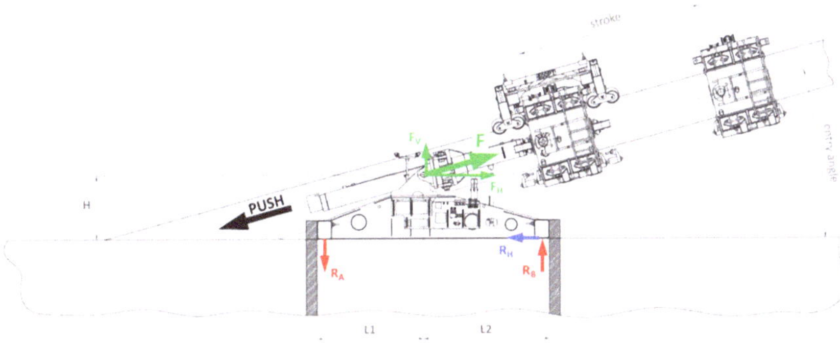

Figure 8-1. Pipe thruster free body diagram—thrusting forces. Source: Courtesy of Herrenknecht.

$$R_H = F \times \cos \alpha \tag{8-3}$$

where

R_A = Reaction force at the rear foot of the pipe thruster;
R_B = Reaction force at the front foot of the pipe thruster;
R_H = Lateral reaction force;
F = Thrust force (pull/push), with $F_H = F \times \cos \alpha$ and $F_V = F \times \sin \alpha$; and
α = Entry angle.

When reversing the direction to pull on the pipe string, the reaction forces will reverse. Equations (8-1) to (8-3) will still be valid; their results, however, will switch sign from positive to negative and vice versa. The

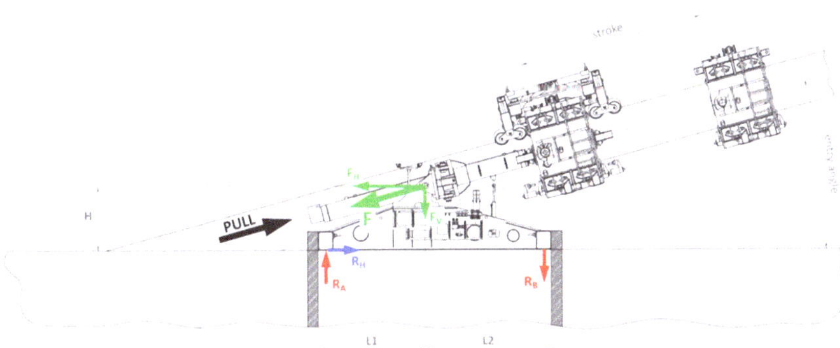

Figure 8-2. Pipe thruster free body diagram—retraction forces. Source: Courtesy of Herrenknecht.

aforementioned equations for reaction forces do not consider additional forces of the weight of the pipeline and the pipe thruster's own weight. Because they are for general guidance, contractors and engineers should verify thruster geometry and reaction forces with the respective equipment manufacturers.

8.4.5.2 Deep Foundations. The magnitude of the reaction forces involved in DSPT installations often require temporary deep foundations to provide the necessary support to the pipe thruster for the DSPT operations. Deep foundations may be driven or drilled and installed vertically or on a batter, as required by the subsurface conditions. If dense or hard ground conditions are present near the ground surface, shallow or drilled foundations may be suitable to resist reaction forces. Typically, the lateral reaction and uplift forces govern the foundation requirements.

8.4.5.3 Thrust Block Foundations. Shallow foundations may be used for short drives with shallow angles where the anticipated thrust loads are low and mainly consist of lateral forces. Alternatively, dead-weight anchoring of the pipe thruster can be used by adding weight to the base of the pipe thruster.

8.4.5.4 Working Pad. The DSPT contractor may use a concrete slab, timber mats, or a gravel base for the working pad for the pipe thruster. A concrete pad is often used, and it can serve as a structural cap for deep foundations.

8.4.5.5 Pipe Thruster Setup and Anchoring. The pipe thruster should be properly anchored to the foundation built for DSPT operation. Depending on the foundation involved and contractor preferences, the anchoring system typically includes a structural steel frame. Figure 8-3 provides a general illustration of a pipe thruster setup.

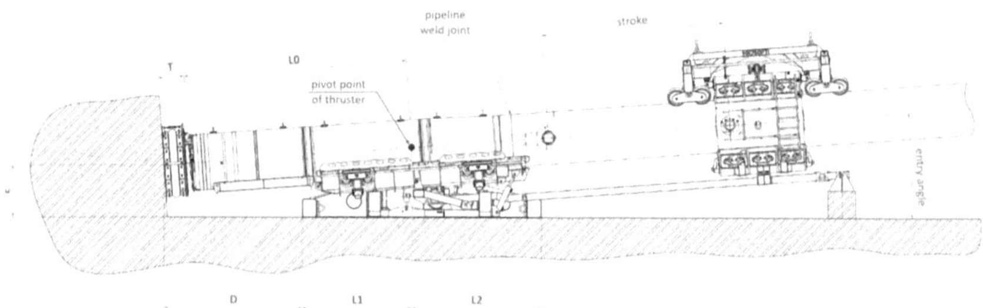

Figure 8-3. Pipe thruster—entry point distance for a DSPT launch.
Source: Courtesy of Herrenknecht.

The following equations provide typical setup geometry values.

$$D = ((L0 + T)\cos\alpha) - L1 \tag{8-4}$$

$$E = H - ((L0 + T)\sin\alpha) \tag{8-5}$$

where

D = Horizontal distance from the entry point to the front end of the thruster,
E = Vertical distance from the launch working surface to the entry point (centerline of the launch seal),
T = Thickness of the launch seal,
$L0$ = Distance from the pivot point of the thruster to the launch seal, and
$L1$ = Horizontal distance of the pivot point to the front end of the thruster.

8.4.6 Thrust Pipe Staging, Fabrication, and Testing

Within the available pipe staging workspace, the contractor will assemble, weld, coat, and test one or more string(s) of the thrust pipe. The contractor may want the first section of the thrust pipe to be a short string for the purpose of optical control survey. If the available workspace does not accommodate the entire length of the thrust pipe, one or more intermediate weld(s) during installation will be necessary.

The width required for the installation of DSPT depends on the pipe diameter and number of welds required during the installation process. Enough width should be provided to facilitate fabrication and storage of the required number of pipe strings and to mobilize equipment required to move the pipe strings before the intermediate welds. Adequate width to stage the drilling fluid cleaning system within the pipe stringing area should be considered. If the trailing power-pack module is used, adequate space is needed for it to travel.

The overbend height required during a DSPT installation depends on the entry angle, the ground surface topography, and the allowable bending radius on the thrust pipe. The use of a launch shaft/pit will reduce the required overbend height compared to a surface launch. While estimating the overbend height, the contractor should consider that the bending radius of the thrust pipe starts at the pipe thruster's fully extended clamp position and not at the entry point, as illustrated in Figure 8-4. The section between the clamp and the launch seal should be straight and aligned to the entry angle.

The support requirements depend on the maximum allowable span of the pipeline. When calculating the maximum allowable pipe span, the

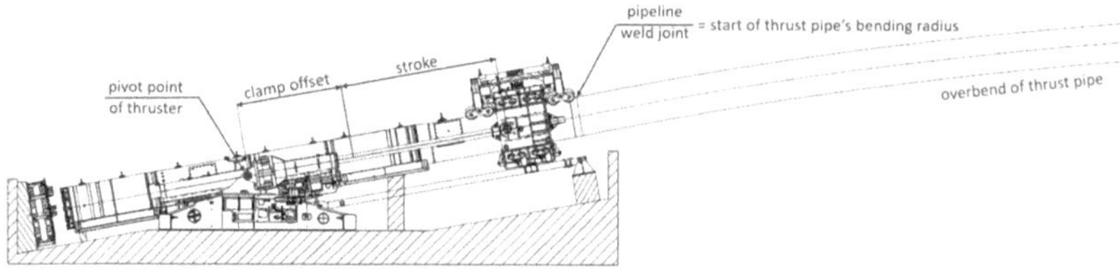

Figure 8-4. Pipe thruster—starting point of an overbend.
Source: Courtesy of Herrenknecht.

weight of the umbilical inside the thrust pipe with flow of both fluid and discharge line should be included.

A temporary half-pipe welded to the tail end of the thrust pipe is used to carry the discharge pump and facilitate umbilical makeup and personnel entry at the tail end of the pipe string, as shown in Chapter 2, Figure 2-10. Adequate space should be provided at the tail end of the thrust pipe section for the half pipe. Typically, the contractor loads the umbilical system from the tail end of the pipe and pushes it toward the front end. The front end of the umbilical must be set so that it lines up with the MTBM connections or a previously installed pipe string umbilical.

Although short DSPT installations are often completed with a single pipe string, longer installations may require multiple pipe strings with associated intermediate welds. The intermediate welding process starts with the DSPT contractor disconnecting the umbilical inside the installed string and cutting off the half pipe at the tail end of the installed string. The second string is then placed behind the installed string to prepare for the weld.

Depending on the method of nondestructive testing (NDT) for the intermediate weld, the connection of umbilical between the new string and the already installed string may be carried out before or after the weld is made. If an NDT method such as ultrasonic testing (UT) is used, connections can be made prior to the weld because the umbilical should not interfere with the testing procedure. If the connections are made prior to the weld, the umbilical must be protected from the heat generated by the welding procedure. Welding blankets are typically used to provide this protection. If X-ray is the preferred method for NDT, umbilical connections are made after the acceptance of the weld. In this case, personnel entry is required for completing the connection inside the pipe. The air quality and temperature must be monitored for personnel entry together with all other measures as described in Section 8.1. To reduce personnel entry into the thrust pipe, UT testing should be considered for NDT for intermediate welds during the DSPT operation. Owners may also require pre- and/or post-installation pressure testing of the pipe string(s).

8.4.7 Cold Weather Considerations

A variety of factors should be considered during setup operations if below freezing temperatures are anticipated as they have an adverse impact on human and mechanical efficiencies and can present a health and safety risk. Pump and valve freeze ups may lead to equipment failure and render the entire fluid system inoperable. To prevent this from occurring, heaters and boilers are used and the appropriate personnel to staff this equipment continuously are needed, resulting in cost and time escalation.

8.5 LAUNCH PREPARATION

Prior to launching the MTBM, equipment and operational testing should be completed and verified. A checklist is routinely developed by the contractor in conjunction with the equipment manufacturer to document these activities.

8.5.1 Testing Thruster and Pipe Clamp Systems

The thruster and pipe clamp systems should be tested for functionality prior to launching the MTBM. The pipe alignment should also be visually inspected so that the clamp slides freely over the thrust pipe in the forward and backward motion. Pressures and forces to move the clamp freely should be recorded.

8.5.2 Testing Computer Systems

All computer functions and readouts should be tested and verified prior to launch of the MTBM. This should include all sensors, input/output controls for all mechanical and hydraulic systems, and components providing feedback to the operator. In addition, as part of the testing protocol, the specified data reports should be compiled and operational parameters verified.

8.5.3 Positioning System

Testing of the navigation system should start with verification and documentation that proper servicing has taken place and that the calibration is verified and documented. If used to measure MTBM elevation, the HWL should be checked for leaks and air pockets. An optical survey should be conducted to establish the initial gyroscope position and the elevation of the MTBM. These data should be entered into the guidance system along with the design alignment and profile data.

The measuring wheel used to track installation distance should be properly secured to the launch seal and manually rotated to ensure accurate readings on the guidance program. After the MTBM enters the launch seal, the measuring wheel needs to be adjusted to compensate for the difference in diameter between the MTBM and the thrust pipe to maintain contact with the thrust pipe.

8.5.4 Preparation of Drilling and Lubricating Fluids

The drilling and lubricating fluids should be designed based on the geotechnical conditions anticipated for the project, as noted in Chapter 5, Section 5.3.4. Variations in ground conditions encountered during tunneling may require modifications to the fluid formulations. The drilling fluid should be designed to have the rheologic properties to transport the cuttings from the MTBM to the separation plant and to lubricate cutter tools and other wear parts inside the MTBM slurry circuit.

The lubricating fluid should be designed and mixed separately. The lubricating fluid should promote borehole stability, create a filter cake for the surrounding ground, and reduce friction along the thrust pipe.

8.6 MACHINE OPERATION

8.6.1 Navigation and Guidance System

DSPT systems utilize a navigation system based on a gyroscope for directional control in combination with the HWL for elevation control. The corresponding software package is integrated into the control cabin, although stand-alone units are available. Other alternatives such as electromagnetic field systems based on surface detection have been used in conjunction with the gyroscope and HWL systems.

8.6.2 Control Surveys

Depending on the navigation system used, an initial control survey after installation of the first 100 to 300 ft (30 to 100 m) of the pipeline is typically performed. The position of the first (and possibly subsequent) control surveys depends on the project scope, allowable tolerances, contractor preference, and owner requirements.

Most control surveys are performed through personnel entry inside the installed pipeline. The purpose of the control survey is to verify the location of the MTBM and compare it to the location output by the navigation software in the control cabin. Land-surveying techniques using theodolite or total station are used to determine the actual location of the MTBM at the time of the control survey.

The control survey can be performed without personnel entry only in the entry tangent, before the MTBM enters the first curve. In this case, the surveyor will have a clear line of sight from the tail end of the first pipe string to a prepositioned survey target on the back of the MTBM. Alternatively, the first section of thrust pipe may be comprised of a short piece of sacrificial pipe allowing the contractor to cut an opening into the thrust pipe for the surveyor to see the MTBM with reduced personnel entry into the pipe. Depending on a project's scope and required accuracy, this initial control survey may be sufficient.

After completion of the control survey, the position of the MTBM is updated in the computer system with the survey coordinates and elevation of the gyroscope. Depending on the extent of variance between the surveyed position and the original gyroscope output, the system recommends a drift correction value that the surveyor accepts or adjusts based on past experience and/or anticipated (sideways) drift behavior (e.g., changes in ground conditions).

8.6.3 Evaluating Slurry Returns

Monitoring and evaluating slurry returns is performed during construction for several reasons, including

- Comparing actual to theoretical excavated volumes,
- Evaluating and documenting the ground conditions encountered by the MTBM, and
- Detecting possible equipment damage or other issues.

The engineered drilling fluid (EDF) is evaluated periodically, typically at least once per shift, by sampling the clean fluid in the slurry system and performing rheologic tests such as mud weight, marsh funnel viscosity, plastic viscosity, yield point, and sand content. Slurry return monitoring and evaluation involves observing and sampling the materials removed by the various screens and centrifuge that comprise the separation equipment.

Slurry returns are also monitored to infer the geological conditions encountered. By identifying the type and composition of the excavated material, the contractor and project team can evaluate ground conditions compared to those assumed in design. A number of construction factors can be assessed, including advance rates, thrust forces, slurry and EDF rheological properties, equipment wear and tear, over- or underexcavation, and disposal of excavated material. The contractor may also monitor slurry returns with a magnet to capture metal in the slurry stream, which may be an indication of equipment wear.

8.6.4 Fluid Release to the Surface

Fluid surface release may occur when the hydrostatic pressure of lubrication fluid is greater than the formation limit pressure, or the MTBM is close to the surface when the jetting action or slurry system of the MTBM may cause a release; refer to Chapter 5, Section 5.3.2. Surface releases can occur in the following situations:

- At an uncased launch location, if the ground cover and strength is inadequate to withstand the applied fluid pressures;
- At a reception location when exiting to the ground surface; and
- Near a reception location if the launch point elevation is higher than the reception area ground surface.

Releases along the alignment where the MTBM is at appropriate depth and fluid pressures and properties being properly monitored and managed are typically uncommon in DSPT relative to HDD.

Regardless of the level of risk anticipated, the alignment of DSPT should be visually monitored during construction for fluid surface releases. If drilling or lubrication fluid releases occur, they should be managed or mitigated with environmental controls, vacuum trucks, and other methods in accordance with permit requirements.

8.6.5 Slurry Treatment and Cuttings Disposal

Slurry treatment may be necessary prior to the disposal of cuttings generated during DSPT. The excavated material is initially brought back to the surface in a liquid slurry form. Solids are separated from the liquid to provide greater disposal efficiency. The liquid is then able to be reconditioned and returned to the DSPT process, whereas the solids are properly disposed of. Contractors have a variety of methods for the treatment and solidification of slurry. Often, disposal of the material is better suited in solid form and so the use of desanders, desilters, and centrifugal units are beneficial. The slurry that cannot be solidified is, in general, hauled off in vacuum trucks or disposed of via land farming.

8.7 CONTINGENCY MEASURES

8.7.1 Equipment Failure

Equipment failures can be grouped into two main categories: aboveground and underground. Aboveground equipment failures are related to the pipe thruster, separation equipment, power module, communication cables, or any of the aboveground slurry and lubrication pumps. Equipment failures are mitigated with on-site or readily available replacement parts. These types of equipment failures are often relatively easily addressed and typically do not jeopardize the installed pipe. However, on-site spare parts may be required or considered for construction in remote locations, during extreme weather, or for projects with accelerated schedule requirements.

Many underground failures of equipment components can be addressed using personnel entry into the pipe string to diagnose and replace worn or failed components, assuming that the thrust pipe and MTBM are of sufficient diameter to allow entry. If personnel entry is not possible or repairs cannot be made while the MTBM is in the ground, the pipe thruster will need to be used to retract the pipe string and the MTBM to address the equipment failure at the ground surface. If retraction and in-place MTBM repairs are not possible, a temporary shaft will need to be excavated from the ground surface for external access. During retraction, a volume of lubrication is often added at the front of the MTBM consistent with the rate of retraction to maintain a full column of lubricating fluids to promote hole stability. Filling the excavated hole may not be necessary or feasible in certain ground conditions. Consideration should be given to roadways and any other settlement-sensitive features overlying the alignment in the event retraction is contemplated owing to the risks associated with borehole settlement and collapse. Retraction of the MTBM and pipe string may require cutting the pipe string to allow it to fit into the available stringing area. During any retraction of the pipe string and MTBM, the contractor will need to wash the pipe string of any debris that may cause damage to the pipe where the thruster grips onto it to prevent unnecessary damage to pipe coatings.

During any retraction, the contractor should perform an inspection of the MTBM to determine if any issues need to be addressed prior to advancing back into the excavated bore. If retraction is required, the equipment manufacturer's procedures for retracting the MTBM must be followed to reduce the risk of further equipment damage.

8.7.2 Buried Objects and Obstructions

During the design phase, a thorough geotechnical and site investigation program should be completed to identify the anticipated ground conditions and potential buried objects and to allow the contractor to select an appropriate cutter wheel for the ground conditions. Often the cutter wheel is selected to break down the largest particle with the trade-off of slower production in less-difficult ground conditions. In the event a buried object is encountered that prevents forward progress of the installation, it is considered an obstruction and will either need to be removed or the installation redesigned to avoid it. Removal of an obstruction can be complicated depending on its location and depth. If the obstruction is shallow, an excavation could be completed down to the front of the MTBM to remove the obstruction. This excavation, however, may present challenges associated with the migration of slurry or lubrication fluids through the backfilled excavation. Excavation support may be required to remove the obstruction.

If the depth is too great or the obstruction cannot be removed, the MTBM and pipe string will need to be partially or fully retracted and the alignment and/or profile modified to allow installation around the obstruction. Any redesign of the alignment will need to meet allowable design criteria of the pipe. If a new alignment and/or profile is considered, an investigation should be completed to identify the relative size and composition of the obstruction. Steering out of an existing tunnel can be difficult and typically requires a significant grouting program to mitigate the previously mined tunnel's unsupported condition.

Alternatively, it may be possible to retract the MTBM and pipe string and change the type of cutter wheel to permit mining through the buried object. This approach should include an investigation of the size, composition, and strength of the buried object to determine if a cutter wheel change will effectively excavate through the encountered material. If the buried object is an impassible obstruction, an alternative construction method will be required to complete the crossing.

8.7.3 Soft or Loose Soils

The encountered soils must provide rigidity to support the weight of the MTBM and facilitate steering corrections. Soft or loose soil formations should be identified in the geotechnical investigation conducted during the design phase. If ground conditions lack strength or density, this may restrict the contractor's ability to accurately steer the MTBM along the design alignment and profile. The contractor should closely monitor line and grade during advancement in soft soil formations and cease forward advancement immediately if line and grade cannot be maintained. In this instance, ground improvement or realignment may be necessary to complete the crossing.

In the event soft soil formations are encountered during mining operations, the MTBM operator must react quickly to balance the required face pressure to reduce the risk of overexcavation of soft or loose materials and manage potential sinkholes and ground surface settlement. In extreme situations, if a void or extremely soft or liquefaction-prone soils are encountered, they could make MTBM steering and retraction difficult.

8.7.4 Steering Difficulties

Steering difficulties can occur during an installation depending on the encountered ground conditions. Rock layers or large cobbles/boulders may deflect the MTBM. In soft or loose formations, or in formations containing voids, the MTBM will naturally tend to drift to the path of least resistance. If an unacceptable misalignment is noted, the MTBM can be retracted and the operator can attempt to correct the tunnel deviation. If the contractor is unable to correct the deviation, a redesign of the crossing will be required.

8.7.5 Out-of-Tolerance Mitigation

If the DSPT installation deviates from the tolerances set forth within the contract documents, all operations should cease until the contractor has consulted with the owner's representative, engineer of record, and other stakeholders. If the deviation is accepted, the contractor may resume operations and attempt to correct the deviation as forward progress is made. If the deviation is not acceptable, the contractor will need to attempt to retract the pipe string to a point where correction of the deviation is achieved. Any retraction and associated backfilling of the abandoned tunnel will need to meet the owner's requirements. If correction attempts are unsuccessful, realignment or alternate construction methods need to be considered.

8.7.6 Loss-of-Fluid Circulation

Some amount of drilling fluid loss to the formation is expected, particularly in permeable ground conditions. Fluid circulation within the slurry system should be monitored. If fluid loss is observed, overpressurization of the formation may be the cause, in which case it may be remedied by increasing the flow rates on the return side of the system and decreasing the flow rates on the feed side. Alternately, if the formational fluid loss is manageable, flow rates may be adjusted to maintain cuttings return. In addition, increasing the viscosity or introducing additives to the slurry system may reduce or prevent formational fluid loss. If a complete loss of circulation occurs, the MTBM should be temporarily stopped, the slurry system put into bypass mode to reduce further fluid loss, and the contractor should evaluate options for regaining circulation. These options may include fluid modifications, revising the DSPT alignment and/or profile to more favorable ground, and grouting of subsurface voids.

Annular lubrication fluid volumes should also be monitored for formation losses. Pumping rates of annular lubrication fluid may be decreased or the viscosity increased to mitigate fluid loss and potential inadvertent returns risk until the MTBM is past the zone where fluid loss occurred.

8.7.7 Hole Collapse

Collapse or closure of the annular space outside of the installed pipe during installation will not be detected until jacking forces increase or noticeable settlement is observed at the ground surface. Proper use of annular bentonite lubrication fluid with sufficient volumes and properties to hold the annular space open is ordinarily effective at preventing collapses during installation. If collapse of the annular space occurs, which inhibits the ability of the contractor to proceed with DSPT, corrective action could include retraction of the entire pipe string and grouting of the excavated hole.

A common hole collapse scenario occurs during retraction of the MTBM to perform extensive maintenance or tooling changes. During this process, the excavated hole is supported by the drilling fluid and full or partial collapse is a risk. If hole collapse occurs, the surface should be closely monitored for settlement. The area above the alignment should be cordoned off until remediation measures are implemented. When minor ground settlement occurs, depending on the settlement impacts, it may be mitigated after DSPT is complete.

If the MTBM is retracted and the resulting excavated hole is abandoned, more permanent means of backfilling the excavation, such as cement grout, should be considered.

8.7.8 Pipe-Coating Damage

If pipe-coating damage occurs, and is isolated to a specific section, the thrust pipe can be cycled through the bored hole using the pipe thruster to advance additional sections of product pipe. If the coating damage occurs from the pipe thruster equipment or launch casing, this can easily be detected, and corrective action can be taken prior to the pipe being advanced past the entry seal. Unfortunately, as in most trenchless applications, certain ground conditions may cause coating damage. Additional coating or a larger overcut should be considered if coating damage is considered a risk.

8.8 CONSTRUCTION DOCUMENTATION AND AS-BUILT SUBMITTALS

Detailed documentation of construction processes, installation loads, geometry, and other items are critical to confirm that design and permit requirements were fulfilled and that it is appropriate for the owner to accept the installation and place the installed pipe into operation. Good documentation reduces uncertainty for both the owner and the contractor with respect to the change in condition claims and other potential construction disputes.

8.8.1 Owner Construction Observation

Detailed documentation and timely construction data reduction and analysis are critical to the successful and safe completion of a DSPT crossing. These services are routinely provided by the engineer of record or a third-party engineer acting as the owner's representative. During construction, the horizonal alignment and vertical profile of the MTBM should be documented along with the thrust forces required to complete the drive. If annular pressure is a permitting requirement and/or critical design factor as with some levee crossings, it should be monitored and documented to confirm that the annular pressures related to the minimum factor of safety against hydraulic fracture are not exceeded. Engineers with a comprehensive understanding of axial, bending, hoop, and combined stresses; buckling; and hydraulic fracture risk should review the construction data and provide timely input into the construction process to reduce the risk of steel pipe yielding issues, hydraulic fracture of the ground, or other undesirable outcomes (Robison and Arens 2017).

8.8.2 As-Built Drawings

On completion of the DSPT installation, the contractor is often required to produce an as-built drawing of the installed thrust pipe. This should occur prior to owner acceptance. The engineer of record should review the as-built data and provide feedback to the owner regarding the conformance of the installation to design tolerances and other requirements.

8.8.3 Construction and Survey Data

Construction and survey data are typically obtained, documented, and submitted by the contractor and may be independently obtained or verified by the owner's engineer during DSPT construction operations. Depending on contract requirements, the following items (among others) may be documented and reported on a per-shift basis:

- Daily weather conditions;
- Shift production and percentage of work completed to date;
- Steering data, horizontal and vertical deviations, if any, from the design plan and profile;
- Thrust force, face pressure, and penetration rates;
- Annular pressure (if required);
- Drilling and slurry fluid testing reports;
- Schedule analysis;
- Material/supplies delivered to the site;
- Type of equipment in use;

- Size of workforce, supervisors, client representatives, and inspectors;
- Hours worked and downtime per day;
- Areas of concern and corrective actions taken (if any);
- Photographic documentation; and
- Daily safety meeting documentation.

Detailed documentation of these items, and any others specifically required contractually, provides for a significant reduction in uncertainties in the project record. Building a clear and detailed project record reduces the likelihood and scale of owner–contractor disputes.

REFERENCES

Carpenter, B., M. Carlin, and A. Snider. 2018. "An overview of safety considerations for large diameter Direct Pipe® installation." In *NASTT 2018 No-Dig Conf.* Paper TM2-T5-01. Bothell, WA: North American Society for Trenchless Technologies.

Robison, J., and N. Arens. 2017. "Direct Pipe® construction—An engineering perspective on quality assurance." In *Pipelines 2017: Planning and Design*, edited by A. Pridmore and J. Geisbush, 227–238. New York: ASCE.

GLOSSARY

ADR: Alternative Dispute Resolution—A dispute resolution process that acts as a means for disagreeing parties to come to an agreement short of litigation.

Advance Rate: Forward progress over a given period, which includes penetration rate, makeup time, down time, and idle time; typically measured in feet/shift or meters/shift.

Annular Space: Volume between the gauge cut (surrounding ground left in place) and the outside radius of the machine or body of the thrust pipe multiplied by the length of the installation.

Annulus: Area between the gauge cut and the pipe outside diameter (OD) equal to (Annulus area = $\pi/4$(gauge cut OD^2 − pipe OD^2)).

API (American Petroleum Institute): National trade association representing the oil and gas industry; members include large integrated companies, as well as exploration and production, refining, marketing, pipeline and marine businesses, and service and supply firms.

ARO (Abrasion Resistant Overcoating): Secondary layer of coating for FBE-coated pipes that is designed to protect the primary FBE coat from impact and/or damage.

Bar: Unit of pressure defined as the average atmospheric pressure at sea level.

Bentonite: Expansive clay that forms a slick slurry or gel when water is added.

BHA: *See* Borehole Assembly.

BMBF: German Federal Ministry of Education and Research (Bundesministerium für Bildung und Forschung).

Borehole Assembly: Expression derived from the term "*bottom hole assembly*" (BHA), which is a component of a drilling rig (e.g., HDD or oilfield). Lowest part of the drill string, extending from the drill bit to the drill pipe. The assembly can consist of drill collars, subs such as stabilizers, reamers, shocks, hole openers, and the drill bit sub itself and the drill bit. As it pertains to HDB and HDD, the term "borehole assembly" refers to the drill and cutting head mounted in front of the pipeline.

Boulder: According to ASTM D653, a rock fragment with a principal dimension of 12 in. (305 mm) or more.

Buckling: Lateral deflection of a long, slender structural member subjected to an axial compressive load.

Buried Objects: Items in the ground that can affect boring equipment advancement and line and grade control. Buried objects that prevent continued advancement along the design path are obstructions.

Carrier Pipe: Pipe for conveyance of water, gas, sewage, or other products.

Casing: Larger-diameter thrust pipe that eventually houses smaller pipe(s) or conduit(s).

Centerline: Central horizontal axis of the design alignment, pipe, or boring equipment.

Coating: Protective layer applied to the outside of a pipe. *See* ARO and FBE.

Cobble: According to ASTM D653, a rock fragment with principal dimension between 3 and 12 in. (75 and 305 mm).

Cohesionless Soils (Noncohesive Soils): Earth materials containing less than 20% soil particles passing the No. 200 sieve. In addition, the Atterberg limit test must have a plastic index less than 7 and a liquid limit less than 25. Any free-running soil such as sand or gravel whose strength depends on friction between particles.

Cohesive Soils: Earth materials containing 20% or more soil particles passing the No. 200 sieve. In addition, the Atterberg limit test must have a plastic index greater than 7 and a liquid limit greater than 25.

Conical Section: Steel section located behind the MTBM that gradually transitions the outside diameter of the MTBM to the outside diameter of the thrust pipe.

Contract Documents: Documents prepared by the owner and project engineer for bidding and for awarding a project; can include bid forms, general conditions, special conditions, technical specifications, drawings, geotechnical data reports, and geotechnical baseline reports.

Controls: Part of the DSPT system that allows synchronized excavation, removal of spoils, and installation of pipe while simultaneously supporting the ground and counterbalancing hydrostatic pressure to balance forward movement with excavation so that ground settlement or heave is minimized.

Cover: *See* Ground Cover.

Crossing: Pipeline installation in which the primary purpose is to provide one or more passages beneath a waterway, road, or other surface obstructions.

Cutter Head (Wheel): Any rotating tool or system of tools on a common support that excavates at the face of a bore.

Desktop Study: Compilation and review of a publicly available site, geotechnical, and other information pertinent to a project. Typically performed at the outset of a project, prior to field explorations and detailed design.

Dewatering: Act of removing groundwater or lowering the groundwater elevation using a system of wells and pumps.

Differing Site Condition Clause: Clause included in the contract documents to address how unanticipated ground conditions are handled.

Differing Site Condition (DSC): Physical condition encountered in the tunnel alignment that differs materially from reasonably anticipated conditions.

Direct Pipe: Trenchless method of pipeline installation developed and patented by Herrenknecht, AG, in Germany in the early 2000s. Direct Pipe combines aspects of microtunneling and horizontal directional drilling, utilizing the innovation of the pipe thruster to provide the required force to advance (or retract) the MTBM, trailing cans, and thrust pipe. The force is applied to the outside of pre-assembled steel pipes. Direct Pipe is included in the broader category identified in this manual as direct steerable pipe thrusting (DSPT).

Downtime: Time lost when DSPT equipment is unable to operate.

Drilling Fluid: Water that may contain additives such as bentonite, polymers, soda ash, surfactants, or other materials to enhance face stability, ground excavation, and cuttings-carrying capacity. If the drilling fluid contains additives and is designed to have specific properties, it is considered an engineered drilling fluid (EDF).

Drive: Thrust pipe installation from the launch to the reception points.

Dry Recovery: Retrieval of the MTBM by removing it from a dry, land-based pit or shaft or from a dry, watertight shaft or cofferdam in a water body.

DSPT (Direct Steerable Pipe Thrusting): Category of near-surface launched, thrusted microtunnel methods of pipeline installation that install steel pipes in a single pass, often along horizontal and vertical curves; includes Direct Pipe and System2.

Engineered Drilling Fluid (EDF): *See* Drilling Fluid.

Entry Seal: *See* Launch Seal.

Entry Wall: Defined in this book as the wall or structure inside the launch shaft that secures the launch/entry seal through which the MTBM enters the ground.

Exit Pit: *See* Receiving Pit.

Face: Location in front of the MTBM where excavation occurs.

Face Pressure: Earth and groundwater pressures applied against the cross-sectional area of the face.

FBE: Fusion-bonded epoxy (coating) typically factory-applied to a pipe exterior using a thermal bonding process.

Filter Cake: Thin layer of clay or polymer coating the ground at the face and annular space from drilling and lubrication fluids.

Frac-Out: *See* Inadvertent Returns.

Gauge Cut: Outermost projection of the excavation measured as a diameter.

Gauge Cutter: Outermost cutter that creates the gauge cut (i.e., the nominal size of the borehole) and is subject to maximum wear.

Gel: According to ASTM D653, the condition in which a liquid begins to exhibit measurable shear strength.

Golden Weld: Pipeline weld that connects thrust pipe segments that cannot be pressure-tested before being installed underground. Typically certified by other methods (see NDE and/or NDT).

Ground Cover: Minimum distance (separation) between the ground surface and top of the pipe.

Groundwater: Subsurface water as distinct from surface water.

GRP: Pipe coating consisting of glass-fiber-reinforced plastic.

Guidance System: *See* Navigation System.

Half-Pipe: Pipe section cut longitudinally, connected to the end of the thrust pipe, and used to support, assemble, and organize umbilicals before inserting them in the thrust pipe.

HDPE: High-density polyethylene (plastic pipe).

Heading: Location at the face of the MTBM where excavation occurs and the direction of the drive.

Heave: Upward ground movement.

Hydraulic Fracture: Tensile or shear failure of ground surrounding the annulus or at the MTBM face, typically the result of fluid pressures higher than the strength of the ground. Sometimes results in an inadvertent return.

Hydrostatic Water Level: Water-filled pipe or tubing used to measure the depth of the MTBM relative to a fixed reference point at the launch location.

Inadvertent Returns: Loss of drilling fluid, slurry, or lubrication fluid from the face or annular space to the ground surface or waterway. Sometimes referred to as a *frac-out*.

Intermediate Weld: *See* Golden Weld.

ISO (International Organization for Standardization): Independent, nongovernmental international standard–setting body composed of representatives from various national standards organizations.

Launch Casing: Steel pipe casing located at the launch point through which the MTBM and thrust pipe are passed; typically secures the launch seal and provides support to the surrounding ground.

Launch Pit: Excavation with at least one sloped side in which the pipe thruster is located.

Launch Point: Location where the MTBM enters the ground in the shaft or the launch casing. Alternatively, may be defined as the location where the DSPT geometry intersects the ground surface.

Launch Seal: Mechanical seal, usually comprised of an adjustable steel frame with one or more rubber gaskets that is mounted to the entry wall of the launch pit/shaft or launch casing. The rubber gaskets are distended by the MTBM and thrust pipe as they pass through. The seal is intended to prevent groundwater, soils, slurry, and lubrication fluid inflow into the launch pit/shaft.

Launch Shaft: Fully shored excavation in which the pipe thruster is located.

Lining: Protective layer applied to the inside of a pipe.

Lubrication Fluid: Water–bentonite mixture that may contain polymers pumped into the annular space to support the surrounding ground and reduce the skin friction on the thrust pipe during installation.

Lubrication Ring: Steel ring with lubrication ports, mounted at the back of the MTBM that distributes lubrication fluid into the annular space as the machine advances.

Marsh Funnel Viscosity: Measure of the flow rate of a fluid through a calibrated funnel (Marsh Funnel) in seconds per quart of fluid. The Marsh Funnel Viscosity of water is 26 s per quart.

Measuring Wheel: Calibrated measuring device that rolls on top of the thrust pipe as it is inserted into the ground to provide a measurement of the length of thrust pipe, MTBM, and trailing cans in the ground; typically calibrated to the location of the gyroscopic guidance system.

Mixed Face: Condition of encountering two or more different geologic materials in the face with contrasting engineering properties (e.g., rock overlain by soft ground or very soft, low-strength soil overlain or underlain by a very stiff, high-strength soil).

Mixed Ground: Excavated ground, without a defined interface, that contains soil with fine and coarse fractions of gravels, cobbles, and boulders.

Mixed Reach: Significant change in ground type, characteristics, or behavior in the face between the launch and reception points.

MTBM: Microtunnel tunnel boring machine.

Navigation System: Position controls for DSPT that allow underground navigation and typically include with increasing sophistication: (1) a water hose leveling system for vertical control, (2) a gyroscopic device for determining horizontal location, (3) inclinometers and roll sensors for orientation, and (4) a measuring wheel to measure thrust pipe installation distance.

NDE (Nondestructive Examination): *See* NDT.

NDT (Nondestructive Testing): Umbrella term encompassing several methods for completed weld inspection that do not permanently change the weld.

***N*-Value:** Number of blows per foot required to drive a standard penetration test (SPT) soil sampler into the ground 12 in. (0.3 m) during geotechnical exploration (ASTM D1586). Often used to estimate in situ consistency of clays and silts and relative density of sands.

Obstruction: Buried object that lies completely or partially within the face that prevents continued forward progress along the design path.

One-Pass: Construction method by which the thrust pipe is directly installed in the ground.

Open Cut: Excavation method made from the ground surface to achieve a required depth for the installation or repair of utilities. The excavation is then backfilled and the surface reinstated.

Overbend: Transition of the thrust pipe from a horizontal resting position into a bend that matches the angle of the launch tangent. Configuration of the overbend that is controlled by the pipe's minimum allowable bend radius.

Overburden: *See* Ground Cover.

Overcut: Difference in diameter or radius between the gauge cut and the thrust pipe.

PE: Polyethylene (pipe or coating).

Penetration Rate: Instantaneous excavation distance per unit time, typically measured in inches per minutes or millimeters per minute.

Performance Specification: Document that establishes the performance criteria that the constructed product must meet. Performance specification that leaves the selection of the means and methods to the contractor.

Personnel Entry: Description of any inspection, construction, renewal, or repair process that requires a worker to enter a pipe, duct, tunnel, or bore.

Piezometer: Specialized monitoring well that provides discrete screens and underground seals to measure the groundwater level at specific intervals belowground. The groundwater level may be measured using open standpipes or with vibrating wire transducers.

Pipe Pusher: *See* Pipe Thruster; also, the product name for Prime Drilling, GmbH.

Pipe String: Individual pipe segments joined into longer sections, typically located behind the pipe thruster; becomes thrust pipe as it is advanced beyond the pipe thruster.

Pipe Thruster: Device that grips an installation pipe around the outside and advances it forward (or backward) continuously or intermittently in strokes. Best suited for pipeline installations of long pipe sections; also, the product name for Herrenknecht, AG.

Pitch: Upward or downward angle of the MTBM measured from a horizontal plane passing through the longitudinal axis of the MTBM.

Port: Opening in the MTBM, trailing cans, or thrust pipe to mount sensors or introduce fluids into the annular space.

PP: Polypropylene (coating).

Product Pipe: *See* Carrier Pipe.

PVC: Polyvinyl chloride (pipe).

RCP: Reinforced concrete pipe.

Receiving Pit: *See* Reception Pit.

Reception Pit: Excavation with at least one sloped side into which the MTBM and trailing cans are thrust and recovered.

Rescue Shaft: Unplanned excavation to remove obstructions and/or retrieve or repair the MTBM.

Roll: Angle of rotation about the longitudinal centerline of the MTBM measured from the 12 o'clock (vertical) position.

Settlement: Measurable downward movement of the ground.

Shore Approach: Installation of a pipeline that crosses a shoreline, either from land to waterbody or vice versa.

Sideboom Pipelayer Crane: Track-type or wheel-type tractor having a boom mounted on the side used for lifting, lowering, or transporting a suspended load.

Slurry: Mixture of drilling fluid and spoils.

Slurry Separation: Process in which excavated material is separated from the slurry.

Specification: Contractual document that establishes specific requirements for a product or service.

Spoil: Soil, rock, and other materials excavated during the installation process; also referred to as cuttings or muck.

Stringing Area: Laydown area behind the launch pit used to preassemble the pipe string.

System2: DSPT method produced by Prime Drilling GmbH and mts Microtunneling Systems GmbH utilizing a rack and pinion pipe pusher and a microtunneling MTBM.

Thrust Force: Total force required to overcome the face pressure and frictional resistance along the thrust pipe to allow forward movement of the MTBM and thrust pipe.

Thrust Pipe: *See* Pipe String.

Tooling: Ground-engaging elements of the cutter wheel designed to penetrate and excavate the ground.

Trenching: *See* Open Cut.

Two-Pass: Carrier pipe installed within the thrust pipe or after the DSPT is completed.

Umbilical: Aggregate of all service lines and consists of drilling fluid feed and slurry return lines, lubrication lines, water, compressed air, data communication, guidance, and electric and hydraulic power supply. The umbilical runs through the length of the thrust pipe and typically rests on supports with rollers or casters to protect an interior lined pipeline and facilitate removal of the umbilical.

USACE: United States Army Corps of Engineers.

Yaw: Angle of the MTBM to the left or right of a theoretical vertical plane.

INDEX

FSC
www.fsc.org
MIX
Papier | Fördert gute Waldnutzung
FSC® C083411

Kolibri 360 GmbH
Ferdinand-Jühlke-Straße 7
99095 Erfurt, Deutschland
produktsicherheit@kolibri360.de